ATOMIC STRUCTURE 2

The arrangement of electrons in atoms

Electrons occupy particular energy levels (called shells) in an atom.

Electrons will fill up the lowest (innermost) energy level first.

- The first shell (i.e. the one closest to the nucleus) can hold up to two electrons.
- The outer shells can hold up to eight electrons.

You need to be able to write and draw the electron configurations for the first 20 elements only.

For example: phosphorus (atomic number 15) is 2,8,5.

The electron configuration for sodium (atomic number = 11) can be represented as shown in the diagram.

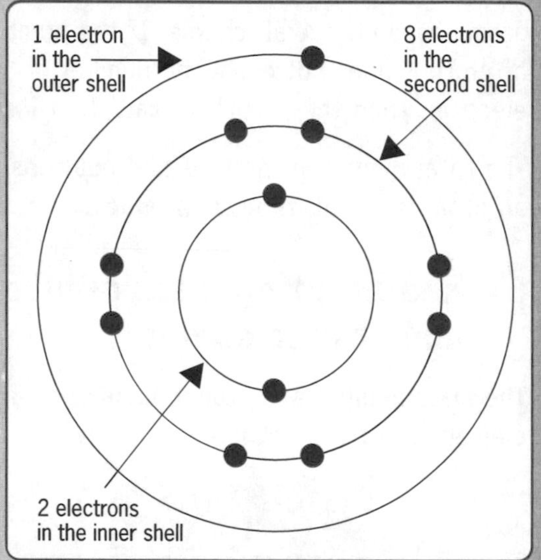

Isotopes

Two atoms of the same element will always have the **same number of protons** but if they have **different numbers of neutrons** they are called **isotopes**.

Chlorine exists as two isotopes.

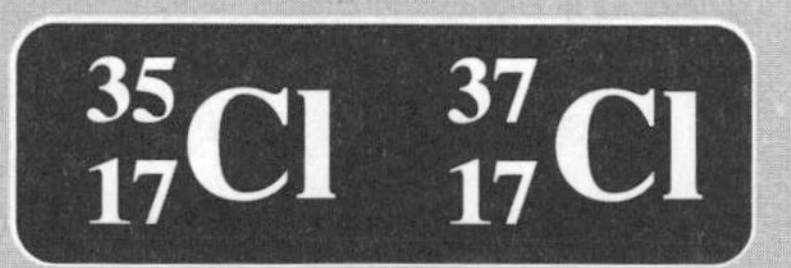

Both types of chlorine atom contain 17 protons (and therefore 17 electrons). They differ because they have different numbers of neutrons in their atoms. One of them (chlorine-35) has 18 neutrons and the other isotope (chlorine-37) has 20 neutrons.

Atoms are the smallest parts of matter. An element is a substance that is made up of atoms with the same numbers of protons and electrons. However, the number of neutrons in an atom can vary.

Atomic number and mass number

The **atomic number** (or proton number) of an atom is the number of protons in an atom. In an atom, the number of protons is equal to the number of electrons, hence atoms have no overall charge. If the number of protons does not equal the number of electrons then the particle is called an **ion**.

The total number of protons and neutrons in an atom is called its **mass number**.

Representing mass number and atomic number

The mass number and atomic number of an element can be represented as follows.

mass number — $^{23}_{11}\mathrm{Na}$ — symbol

atomic number

The **atomic number** of sodium is **11**. The **mass number** of sodium is **23**.

Number of protons in an atom of sodium is **11** (the atomic number of sodium is 11).

Number of electrons in an atom of sodium is **11** (the number of electrons in an atom is always the same as the number of protons).

Number of neutrons in a atom of sodium is **12** (the number of neutrons is the difference between mass number and atomic number).

Progres

1. Which two particles make up the nucleus of an atom?
   a) protons and electrons
   b) protons and neutrons
   c) neutrons and electrons

2. What is the mass of an electron compared to the mass of a proton?

3. What is the charge on a proton?

4. Label the diagram below, showing which number is the atomic number and which is the mass number.

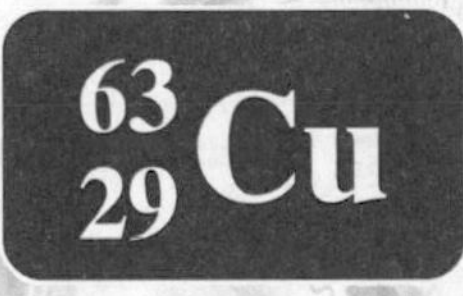

5. Another atom of copper contains 36 neutrons. Show the mass number, atomic number and symbol for this isotope of copper in the same representation as used in question 4.

GCSE IN A WEEK

AUTHOR – DAN EVANS

Use this day-by-day listing and the tabs on each page in the book to plan your revision.

ATOMIC STRUCTURE 1

All substances are made up of atoms.
There are over 100 different types of atom.

Structure of an atom

- The centre of an atom is called the **nucleus**. It contains particles called **protons** and **neutrons**.
- All atoms of a particular element have the same number of protons. Atoms of different elements have different numbers of protons.
- Around the nucleus there are the **electrons**. They are found in different energy levels called **shells**.

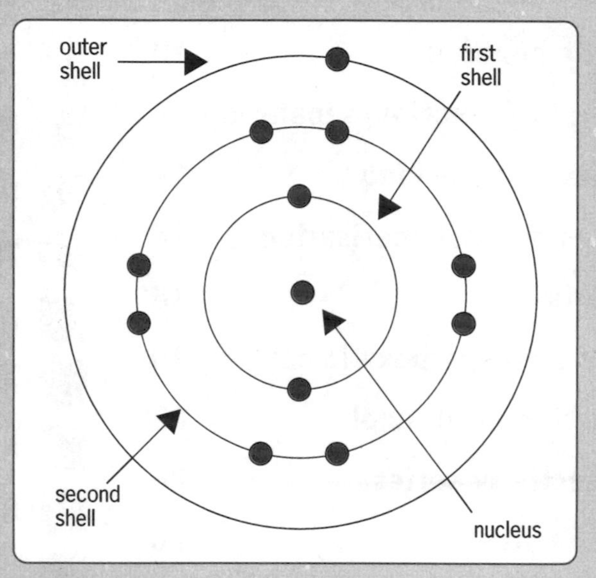

Sub-atomic particles

Protons, neutrons and electrons are collectively known as sub-atomic particles. Their mass and charge are shown in the table.

Particle	Relative mass	Charge
Proton	1	+1
Neutron	1	0
Electron	Negligible	–1

The arrangement of electrons in atoms gives us information about how elements will react. This is very important in explaining chemical reactivity and the arrangement of elements in the periodic table.

Relative atomic mass (RAM)

The relative atomic mass (A_r) of an element is the average mass of the element compared to the mass of an atom of ^{12}C.

This average mass takes into account the masses of different isotopes and their **abundances** (i.e. how much of each isotope there is).

The relative atomic mass is found by multiplying the mass of each isotope by its percentage abundance and then dividing the total by 100.

For example, there are two isotopes of chlorine. Chlorine-35 has an abundance of 75%. Chlorine-37 has an abundance of 25%.

The relative atomic mass of chlorine can be calculated as follows.

$$\text{RAM} = \frac{(\text{mass of isotope 1} \times \text{abundance}) + (\text{mass of isotope 2} \times \text{abundance})}{100}$$

$$\text{RAM} = \frac{(35 \times 75) + (37 \times 25)}{100} = 35.5$$

Hence, the RAM of chlorine is 35.5.

- The fractional value doesn't mean that chlorine has half a neutron! It is just an average mass. If you are asked to work out the number of neutrons in an atom of chlorine, you will be given a specific isotope.

Progress check

1. How many electrons can be found in the first shell in an atom?
2. How many electrons can be found in the outer shells of an atom?
3. Write the electron configuration of sulphur (atomic number 16).
4. Draw the electron arrangement of magnesium (atomic number 12).
5. What is the symbol for the element shown by the following diagram?
6. Define the term 'isotope'.
7. Calculate the relative atomic mass of boron. It consists of two isotopes: boron-10 and boron-11. Boron-10 is 20% abundant and boron-11 is 80% abundant.

IONIC BONDING AND STRUCTURES

Charged atoms are called ions.

Dot-and-cross diagrams

This is a dot-and-cross diagram showing the ionic bond formation in sodium chloride.

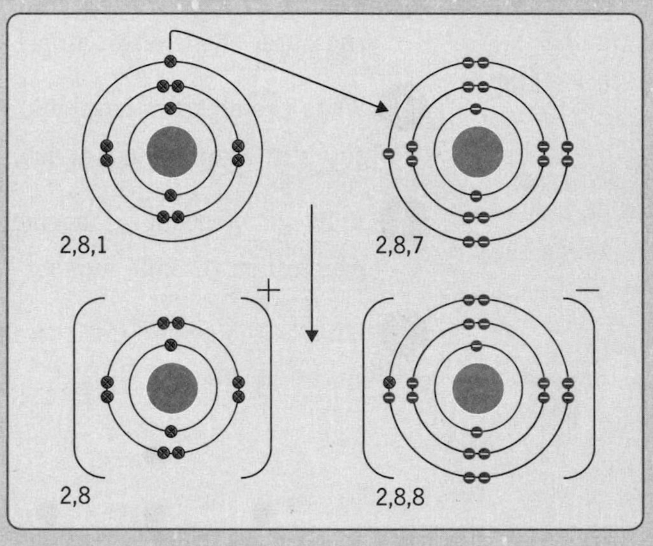

- After the electron has been transferred from the sodium to the chlorine the atoms are charged.
- The ionic bond is the attraction between oppositely charged ions.

For example, this is the ionic bond formation in calcium chloride.

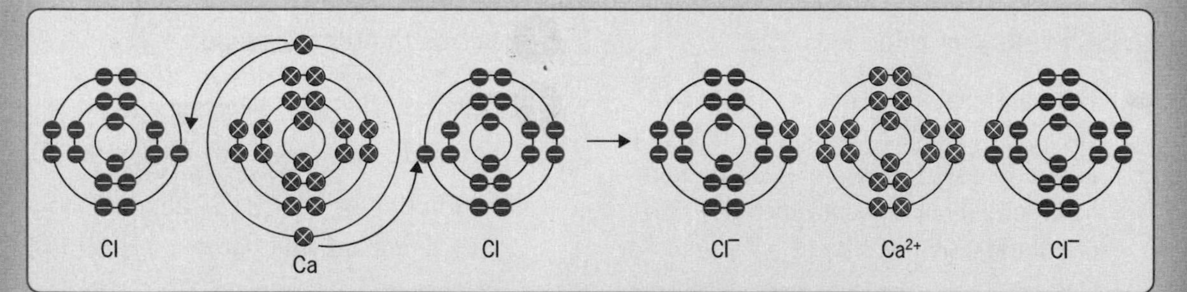

Hence, the formula of calcium chloride is $CaCl_2$.

Ionic bonds occur between metals and non-metals. Electrons are transferred from metal atoms to non-metal atoms. Both atoms end up with a full outer shell of electrons.

Properties of ionic compounds

Ionic compounds form giant structures.

Properties of ionic compounds are shown in the table.

Property	Explanation
High melting and boiling points	Lots of strong bonds
Soluble in water	Water molecules are able to break down the ionic structure
Conduct electricity when molten or aqueous (dissolved in water) but not when solid	Ions need to be free moving in order for electricity to flow Mobile ions act as charge carriers

COVALENT BONDING AND STRUCTURES

DAY 1

Covalent bond formation

The atoms in molecules are held together because they share pairs of electrons. These strong bonds between atoms are called **covalent bonds**. Like ionic bonding, covalent bonding results in the atoms ending up with a **full outer shell of electrons**.

Again, we can use a dot and cross diagram to show the covalent bonds in molecules.

Example 1: hydrogen chloride (HCl)

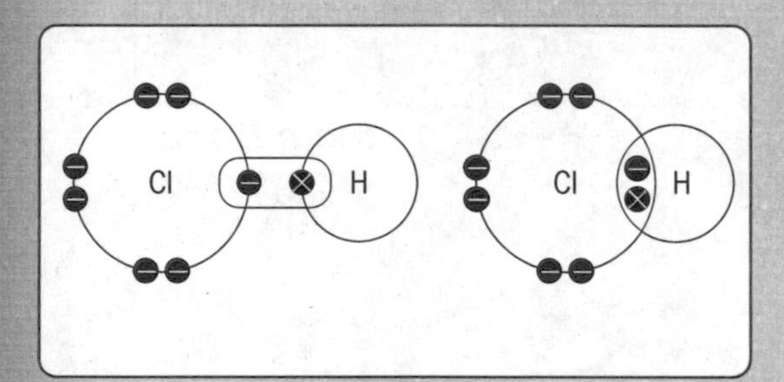

The resultant covalent bond is shown simply as a line between the two atoms.

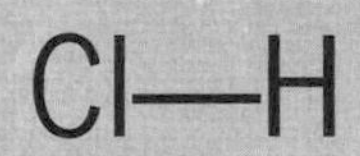

Example 2: methane (CH_4)

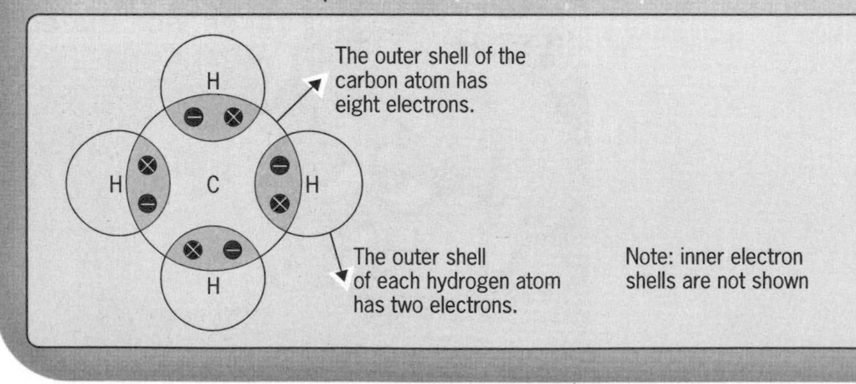

Simple covalent structures

Simple covalent structures are usually either gases (e.g. methane) or liquids (e.g. water). Properties of simple covalent compounds are shown in the table.

Property	Explanation
Relatively low melting and boiling point	Forces between molecules (intermolecular forces) are weak
Do not conduct electricity	The molecules do not carry an overall electric charge

Giant covalent structures

Diamond and graphite are good examples of giant covalent structures.

Both diamond and graphite have lots of strong covalent bonds in their structures. This is why diamond and graphite have very high melting points.

Covalent bonds are formed between two non-metal atoms. The bonds involve the sharing of electrons. There are two types of covalent structure.

In graphite, each carbon atom forms three covalent bonds and the carbon atoms form layers which are free to slide over each other. In graphite there are free electrons on the carbon atoms. This is why graphite can conduct electricity.

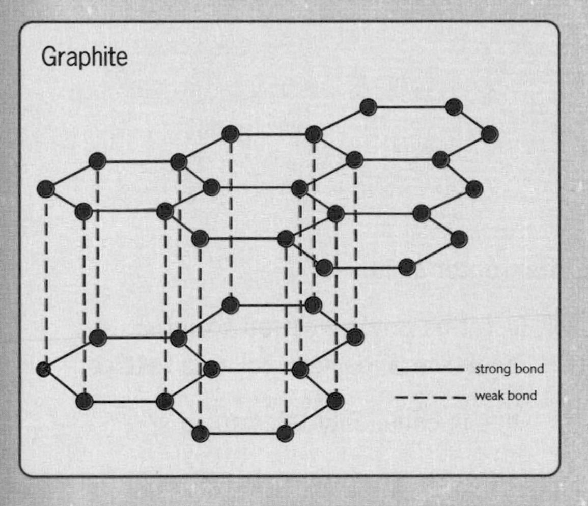

In diamond, each carbon atom forms four covalent bonds in a rigid, giant covalent structure.

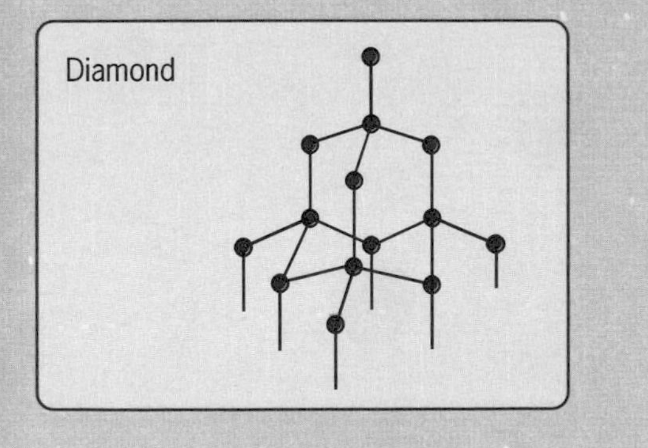

Progress check

1. How many electrons are involved in each covalent bond?

   a) 1
   b) 2
   c) 3
   d) 4

2. What are covalent structures that only have a few atoms in them called?

3. Draw a dot-and-cross diagram to show the covalent bonds in water (H_2O).

4. Why does hydrogen chloride have a low melting point?

5. Why does hydrogen chloride not conduct electricity?

6. Why do diamond and graphite have high boiling points?

METALLIC BONDING

Metal structures

Metals consist of giant structures in which the electrons from the highest occupied (outer) energy levels (shells) of metal atoms are free to move through the whole structure. These free electrons:

- hold the atoms together in a regular structure
- allow the atoms to slide over each other
- allow the metal to conduct heat and electricity

The metallic bonding in sodium

Sodium atoms, for example, have one electron in their outer shell.

In a piece of metallic sodium the sodium atoms will lose this outer electron to form sodium **cations** (Na^+). The cations are arranged in a regular arrangement called a **lattice.**

The electrons move freely between these cations in what is commonly referred to as a **sea of electrons.**

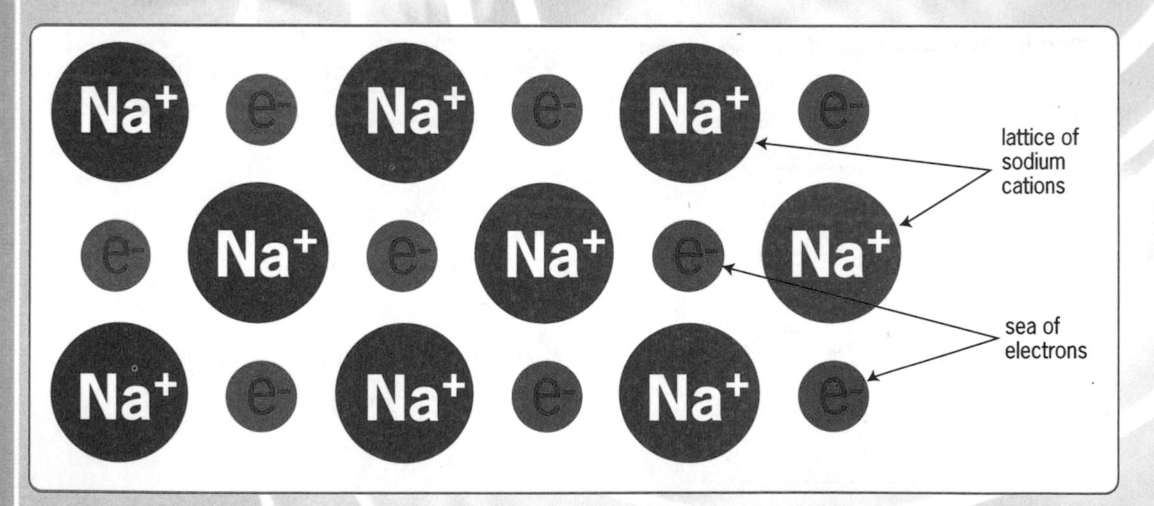

The actual bond is the attraction that exists between the metal cation and the free electrons.

The bonding in metals is similar to ionic bonds in that there is an attraction between oppositely charged particles.

Metallic bonding in magnesium

Magnesium atoms have two electrons in their outer shell. In a piece of magnesium metal the magnesium atoms will have a +2 charge (because each atom will lose two electrons). Hence there will be twice as many electrons as there are cations.

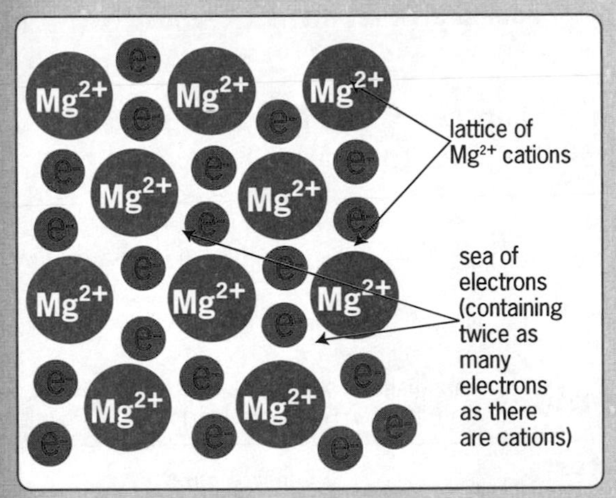

The structure of a metal can be used to explain some of the properties of metals.

Property	Explanation
Conducts electricity	Electrons are free to move throughout the structure
Malleable (bendy) and ductile (can be drawn into a wire)	The electrons can move – preventing the positive cations from getting too close and repelling each other

Progress check

1. What happens to metal atoms when they form metallic bonds?
2. What is meant by the term 'sea of electrons'?
3. Give two properties of metals.
4. Explain why metals can conduct electricity.
5. Draw the structure of calcium metal, showing the lattice of cations and the electrons.

CRUDE OIL: FORMATION AND SEPARATION

Formation of crude oil

The fossil fuels (coal, oil and natural gas) have resulted from the action of heat and pressure over millions of years, in the absence of air, on material from plants and animals (organic material) that has been covered by layers of sedimentary rock.

Crude oil is a mixture of a very large number of compounds.

- A **mixture** consists of two or more elements or compounds that are not chemically bonded together.
- The chemical properties of each substance in the mixture are unchanged. This makes it possible to separate the substances in a mixture by physical methods, including **fractional distillation**.

Most of the compounds in crude oil consist of molecules made of carbon and hydrogen atoms only. These compounds are called **hydrocarbons**.

Fractional distillation of crude oil

The many hydrocarbons in crude oil may be separated into fractions, each of which contains molecules with a similar number of carbon atoms, by a process called **fractional distillation**. This is carried out in a fractionating column. It works by:

1 **evaporating the oil** and

2 allowing it to **condense** at different levels in the fractionating column according to the **boiling point** of the hydrocarbon.

Smaller hydrocarbons travel further up the fractionating column. They have lower boiling points.

This diagram shows the fractions obtained from the fractional distillation of crude oil.

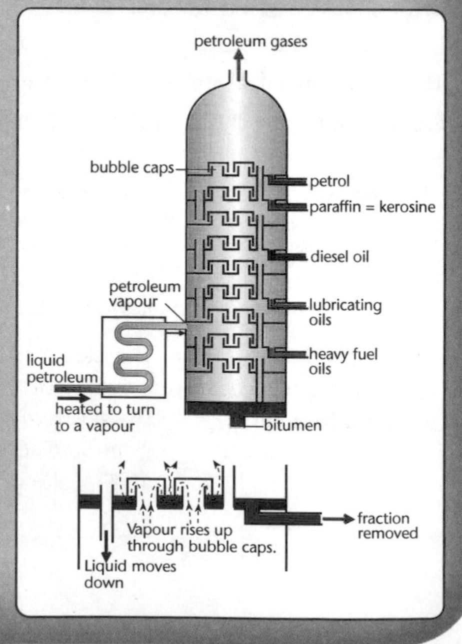

Crude oil is a fossil fuel that is obtained from the Earth's crust. Crude oil is useless on its own. It needs to be separated out into useful materials.

Uses of the fractions

Fraction	Use
Flammable gases	Bottled gas and chemicals
Petrol	Car fuel
Naphtha	Manufacture of chemicals/ pharmaceuticals
Kerosene	Jet fuel
Diesel oil	Diesel fuel
Heavy fuel oils	Fuel for ships
Bitumen	Roads and roofing felt

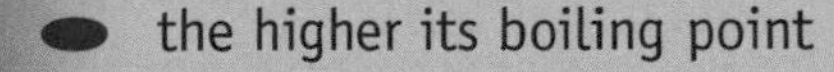

The hydrocarbon molecules in crude oil vary in size. The larger the hydrocarbon molecule (i.e. the greater the number of carbon atoms):

- the higher its boiling point
- the less **volatile** it is (i.e. the less easily it turns into a **vapour**)
- the less easily it flows (i.e. the more **viscous** it is)
- the less easily it **ignites** (i.e. the less **flammable** it is)

This limits the usefulness as fuels of hydrocarbons with large molecules.

Progress check

1. Describe how crude oil is formed.
2. What is a hydrocarbon?
3. How is crude oil separated?
4. On the diagram below, mark an X where the smallest molecules accumulate.

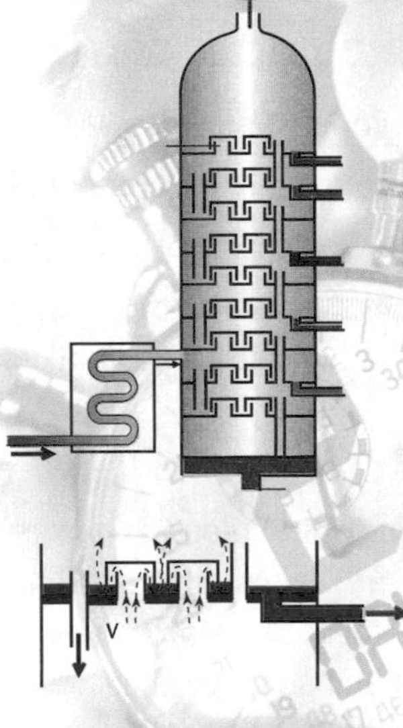

5. By what property are hydrocarbons separated in a fractionating column?

ALKANES AND CRACKING

Alkanes

Alkanes are saturated hydrocarbons.

This means that there are only single bonds in the hydrocarbon molecule. The table shows the formula and structure of the first four alkanes.

Name	Methane	Ethane	Propane	Butane
Formula	CH_4	C_2H_6	C_3H_8	C_4H_{10}
Structure	H—C(H)(H)—H	H—C(H)(H)—C(H)(H)—H	H—C(H)(H)—C(H)(H)—C(H)(H)—H	H—C(H)(H)—C(H)(H)—C(H)(H)—C(H)(H)—H
Number of carbon atoms	1	2	3	4

Homologous series

The alkanes form what is known as a **homologous series.** Members of a homologous series:

- have the same general formula (the general formula of alkanes is C_nH_{2n+2})
- have similar chemical properties
- have physical properties that gradually change as the number of carbon atoms increases (e.g. the boiling point of alkanes gradually increases)

Cracking

Large hydrocarbon molecules can be broken down (**cracked**) to produce smaller, more useful molecules. Some of these molecules are **alkanes** and some of them are **alkenes** (see page 16).

Cracking involves

1 **heating the hydrocarbons** to vaporise them and

2 passing the vapours over a hot **catalyst.**

A **thermal decomposition** reaction then occurs.

This diagram below shows how cracking can be carried out in the laboratory.

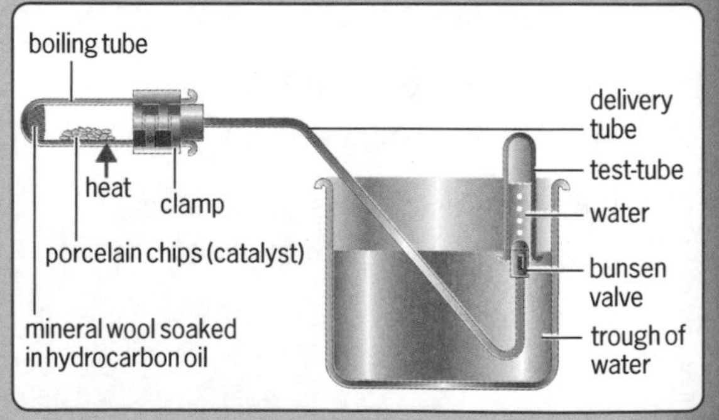

One type of hydrocarbon obtained from crude oil is called an alkane. Small alkane molecules can be used as fuels. Larger alkane molecules can be cracked to make smaller, more useful molecules.

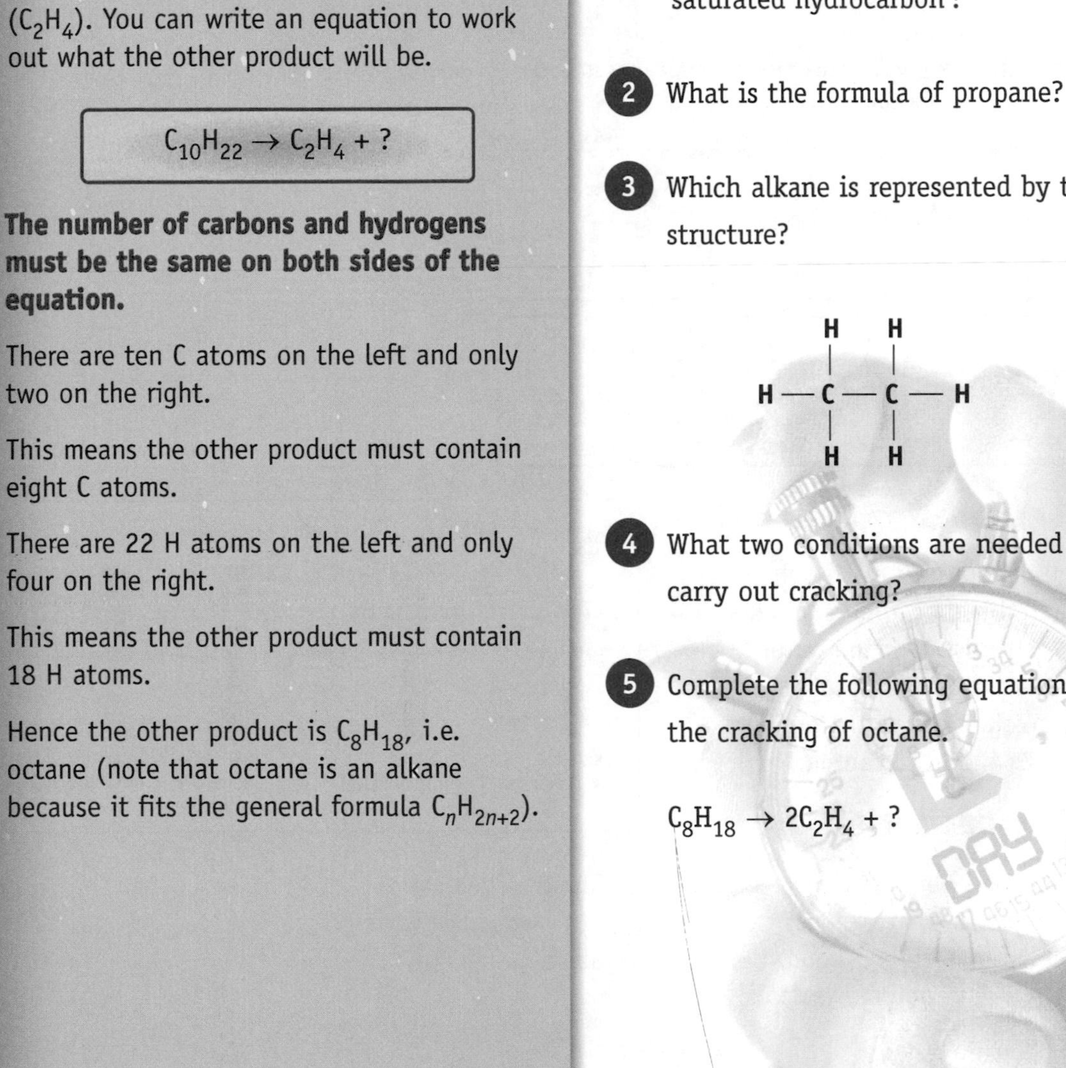

Equations for cracking

If decane ($C_{10}H_{22}$) is cracked then the products could, for example, contain ethene (C_2H_4). You can write an equation to work out what the other product will be.

$$C_{10}H_{22} \rightarrow C_2H_4 + ?$$

The number of carbons and hydrogens must be the same on both sides of the equation.

There are ten C atoms on the left and only two on the right.

This means the other product must contain eight C atoms.

There are 22 H atoms on the left and only four on the right.

This means the other product must contain 18 H atoms.

Hence the other product is C_8H_{18}, i.e. octane (note that octane is an alkane because it fits the general formula C_nH_{2n+2}).

Progress check

1. What do you understand by the term 'saturated hydrocarbon'?
2. What is the formula of propane?
3. Which alkane is represented by this structure?

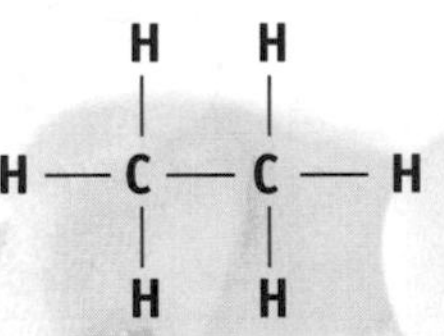

4. What two conditions are needed to carry out cracking?
5. Complete the following equation for the cracking of octane.

$$C_8H_{18} \rightarrow 2C_2H_4 + ?$$

ALKENES AND POLYMERISATION

Alkenes

Alkenes are hydrocarbons that have carbon–carbon double covalent bonds

Alkenes are unsaturated.

These unsaturated hydrocarbons are reactive and so are useful for making many other substances.

The structure of ethene (C_2H_4)

```
H     H
 \   /
  C=C
 /   \
H     H
```

The structure of propene (C_3H_6)

```
H \   / H
H /  C \          / H
      C  =  C
H /            \ H
```

Test for alkenes

A simple laboratory test for an unsaturated hydrocarbon (i.e. an alkene) is to use bromine water. The yellow-brown **bromine water** becomes colourless as the bromine reacts with the hydrocarbon.

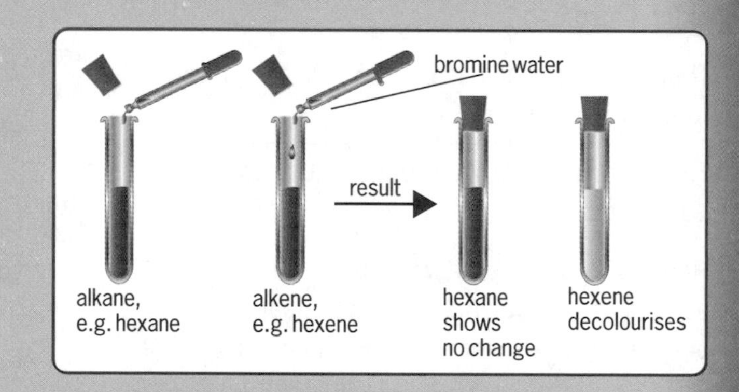

Polymerisation

Polymers have very large molecules and are formed when many small molecules of substances called **monomers** (molecules with a double carbon–carbon bond, i.e. alkenes) join together. This process is called **polymerisation**.

During polymerisation:

- One of the bonds in the double bond breaks.
- A new bond is formed with another monomer molecule.

Alkenes are the more useful hydrocarbon found in crude oil. Their major use is in the manufacture of plastics, a process called polymerisation.

Addition polymerisation occurs when unsaturated monomers join together to form a polymer with no other substance being produced in the reaction.

Plastics are polymers made by polymerisation. For example, poly(ethene) (often called polythene) is made by polymerising the simplest alkene, ethene.

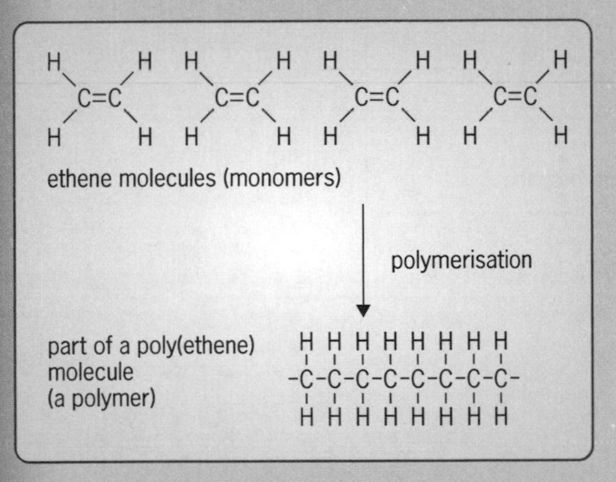

Instead of drawing out a long section of the polymer, you can show the repeating unit.

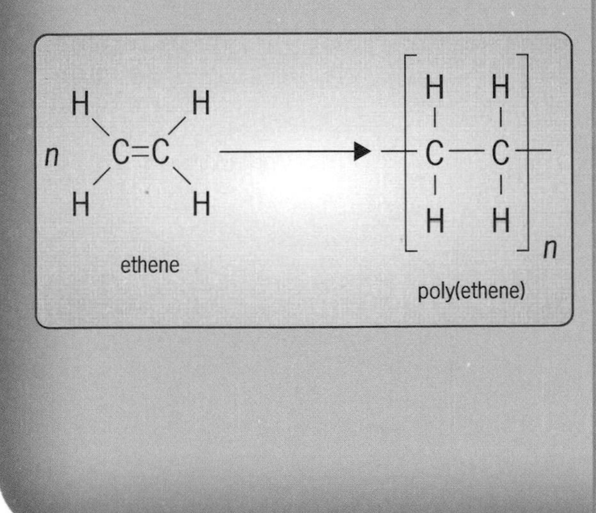

Progress check

1. What do you understand by the term 'unsaturated'?
2. Draw the structural formula of the molecule with the formula C_2H_4.
3. Describe how you could distinguish chemically between a sample of octane (a liquid alkane) and octene (a liquid alkene).
4. What do you understand by the term 'polymerisation'?
5. Draw a section of the polymer formed when propene (C_3H_6) undergoes polymerisation (i.e. show a section of the polymer containing three monomer units).

DAY 2

ISOMERISM

Isomers of butane

Butane has the formula C_4H_{10}. There are two structures that satisfy the formula C_4H_{10}. These are:

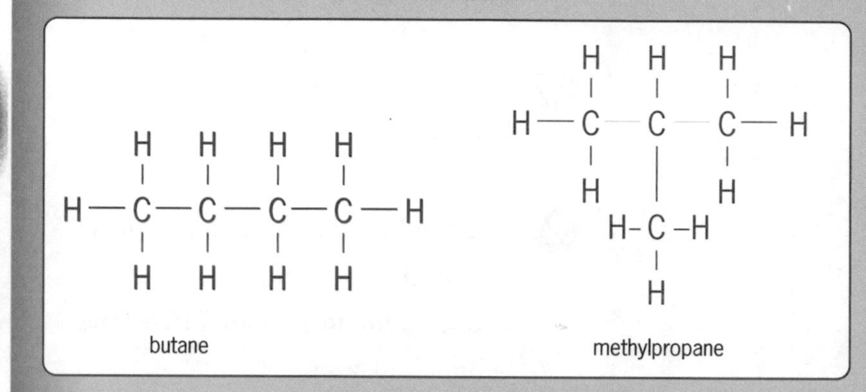

Note that in both these structures each carbon has four bonds and each hydrogen has one bond.

Isomers of pentane (C_5H_{12})

There are three isomers of pentane. They are:

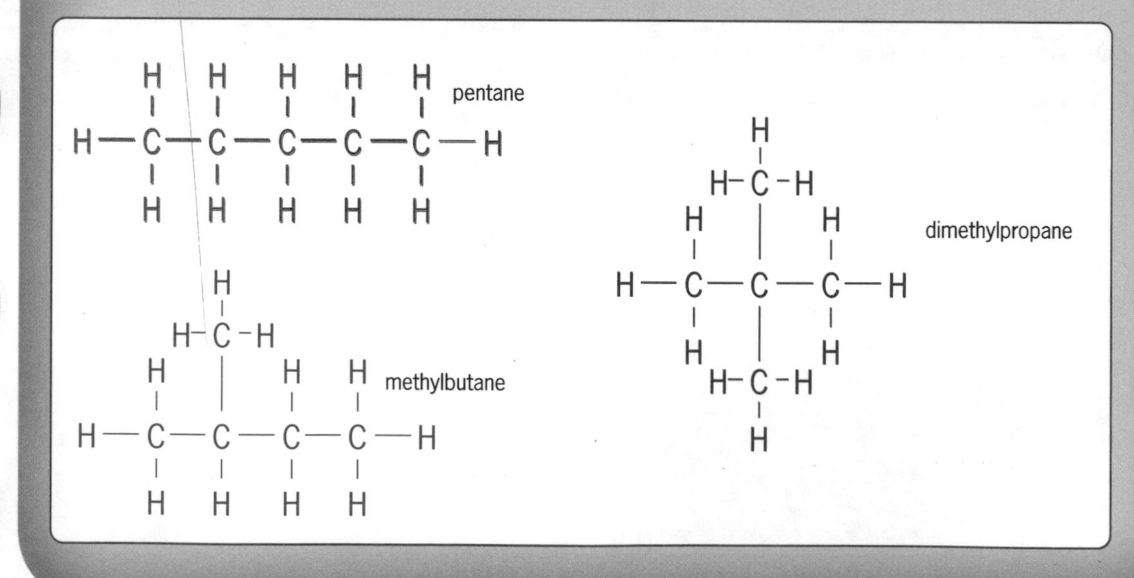

Isomers are two or more compounds with the same chemical formula but that have different structures.

Isomers of alcohols

Alcohols (see page 20) can also exist as isomers. The four isomers of butanol (C_4H_9OH) are:

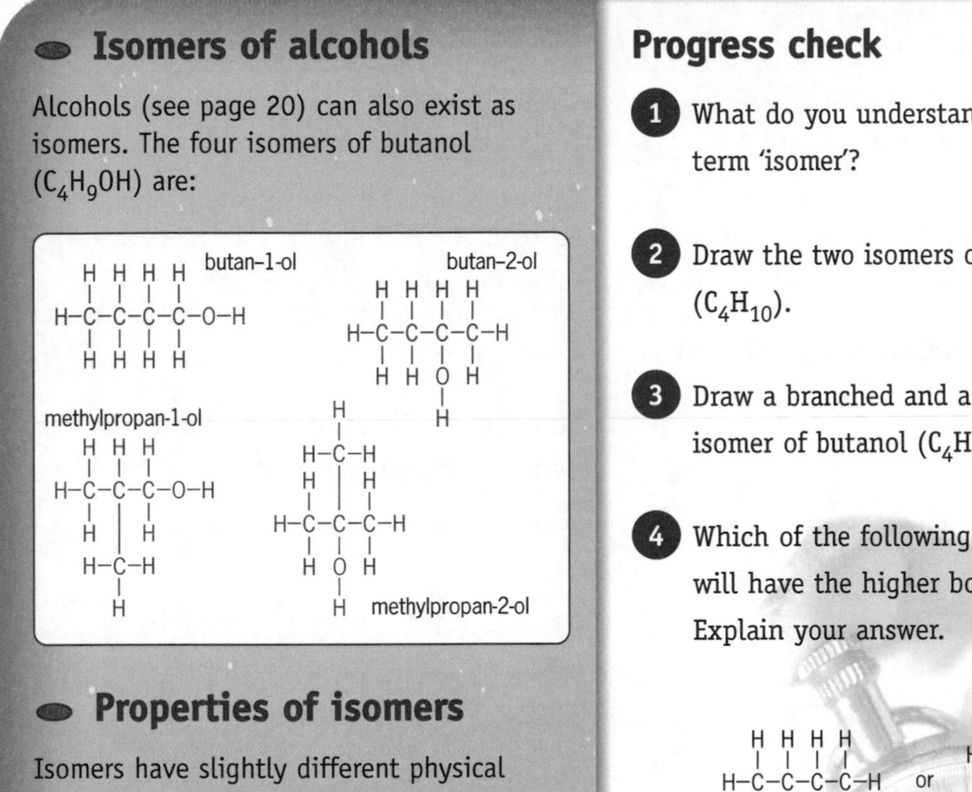

Properties of isomers

Isomers have slightly different physical properties which depend upon the strength of the intermolecular forces (see page 8).

The strength of the intermolecular forces:

- **increases** as the carbon chain length increases
- **decreases** as the amount of chain branching increases

If molecules are able to pack closely together (because of no branching) then the intermolecular forces will be stronger and hence melting and boiling points will be higher.

Progress check

1. What do you understand by the term 'isomer'?
2. Draw the two isomers of butane (C_4H_{10}).
3. Draw a branched and a non-branched isomer of butanol (C_4H_9OH).
4. Which of the following two isomers will have the higher boiling point? Explain your answer.

```
                         H  H  H
   H  H  H  H            |  |  |
   |  |  |  |          H-C--C--C-H
 H-C--C--C--C-H   or     |  |  |
   |  |  |  |            H  |  H
   H  H  H  H             H-C-H
                            |
                            H
```

5. Draw a branched isomer of pentanol ($C_5H_{11}OH$) containing a chain length of four carbon atoms.

ALCOHOLS AND CARBOXYLIC ACIDS

DAY 2

Alcohols

The alcohols form a homologous series with the **general formula $C_nH_{2n+1}OH$.**

It is the OH group (referred to as a **functional group**) that gives alcohols their characteristic properties.

Structures of methanol and ethanol

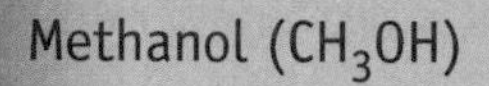
Methanol (CH_3OH)

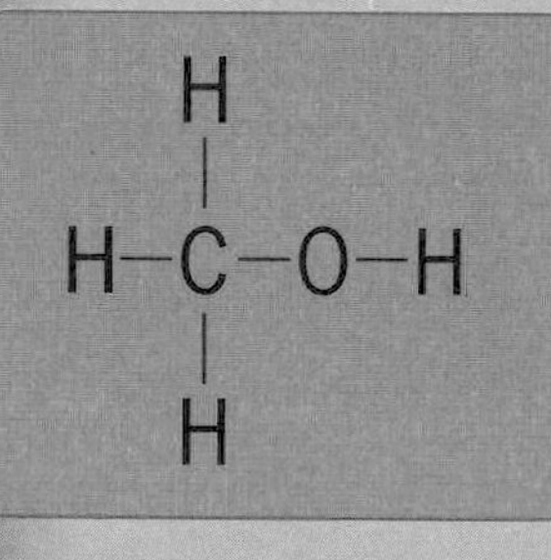

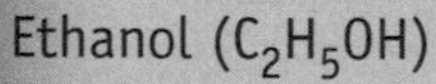
Ethanol (C_2H_5OH)

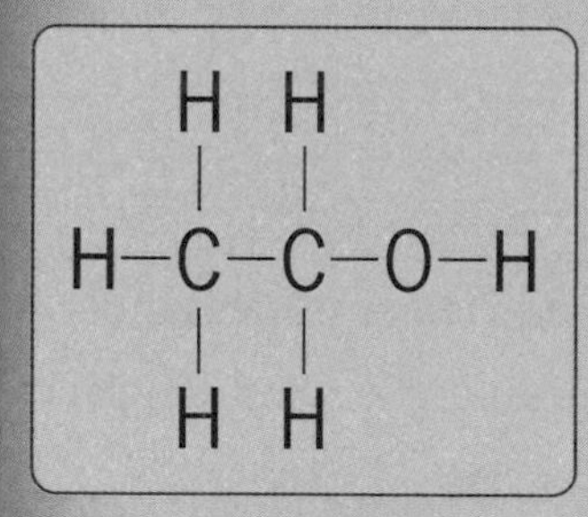

Carboxylic acids

Carboxylic acids form a homologous series all containing the functional group –COOH.

Carboxylic acids are weak acids. They are neutralised by alkalis and they react with carbonates, and hydrogencarbonates to produce salts of the carboxylic acids, carbon dioxide and water.

Structures of methanoic acid, ethanoic acid and propanoic acid

Methanoic acid (HCOOH)

```
    O
   //
H–C
   \
    O–H
```

Ethanoic acid (CH_3COOH)

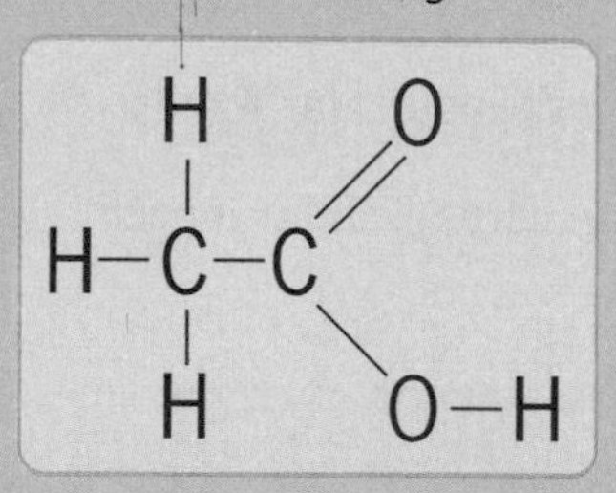

Propanoic acid (C_2H_5COOH)

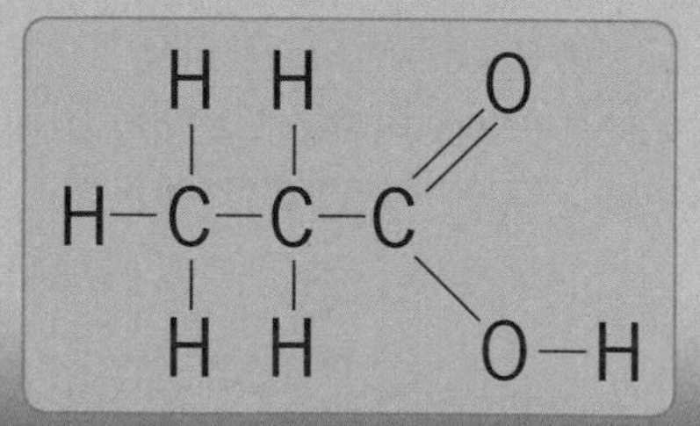

Alcohols are an important industrial chemical. Vinegar and citrus fruits contain molecules known as carboxylic acids, which give the characteristic sharp taste.

Reactions of alcohols and carboxylic acids

Both alcohols and carboxylic acids undergo characteristic reactions.

Reactions of alcohols with carboxylic acids to form esters

Alcohols react with carboxylic acids in a reversible reaction to form esters and water. Esters are widely used as fragrances and food flavourings.

For example, the reaction of ethanol with ethanoic acid forms the ester **ethyl ethanoate.**

Concentrated sulphuric acid is used as a catalyst in this reaction.

Reactions of alcohols with sodium

Alcohols react with sodium metal to form hydrogen gas as one of the products.

Uses of alcohols and carboxylic acids

Substance	Use
Ethanol	Industrial solvent Alcoholic drinks
Cholesterol	Essential steroid to humans
Ethanoic acid	Making esters Making acetate rayon fibres
Citric acid	Found naturally in citrus fruits – used in making drinks
Ascorbic acid	Vitamin C
Aspirin	Pain relief

Progress check

1. What is the functional group present in alcohols?
2. Draw the structural formula of ethanol (C_2H_5OH).
3. Which carboxylic acid is shown by the structure below?

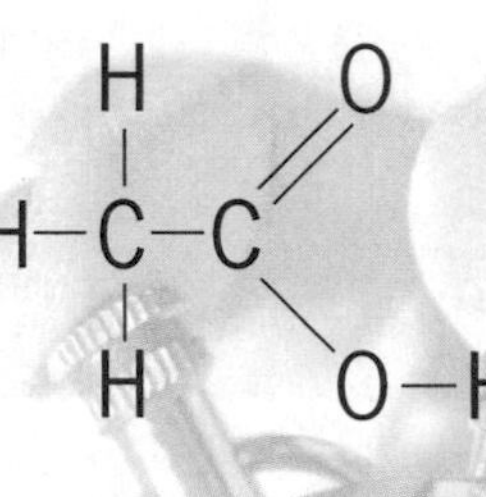

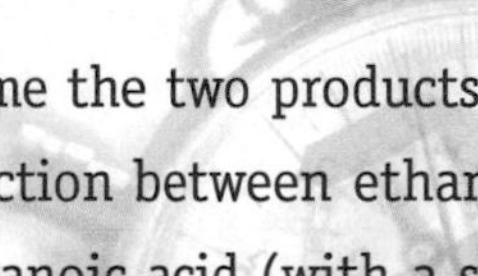

4. Name the two products of the reaction between ethanol and ethanoic acid (with a small amount of concentrated sulphuric acid).
5. What is the function of the concentrated sulphuric acid in the above reaction?

6. Give the use for a named alcohol.
7. Give the use for a named carboxylic acid.

PRODUCTION OF ETHANOL

DAY 2

Fermentation of sugars

Ethanol can be produced by fermentation of sugars.

The raw materials (e.g. fruits containing natural sugars) are mixed with **water** and **yeast** at just above room temperature.

The yeast contains **enzymes**, which are **biological catalysts**.

The sugars react to form ethanol and carbon dioxide.

sugar (e.g. glucose) → ethanol + carbon dioxide

$$C_6H_{12}O_6 \rightarrow 2C_2H_5OH + 2CO_2$$

The carbon dioxide is allowed to escape and air is prevented from entering the reaction vessel.

When the reaction is over the ethanol is separated from the reaction mixture by **fractional distillation**.

Fermentation is a **batch process**.

Diagram of how fermentation can be carried out in a laboratory

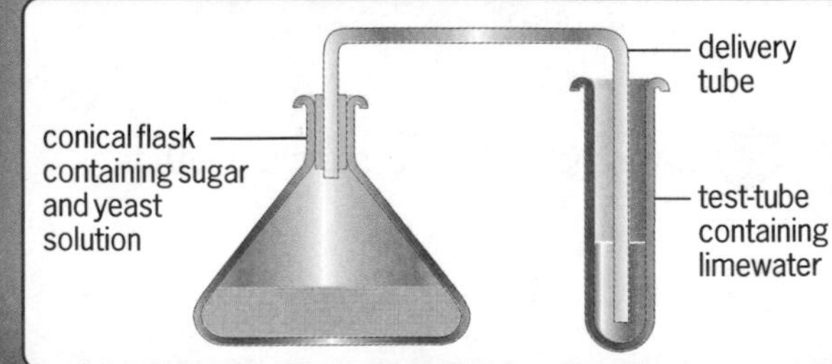

Reaction of steam with ethene

Ethanol can also be made by reacting ethene with steam in the presence of a strong acid catalyst (usually phosphoric acid, H_3PO_4).

The reaction is carried out:

- at a **high temperature**
- at a **high pressure**

This method of ethanol production is a **continuous process**.

Diagram of the continuous production of ethanol

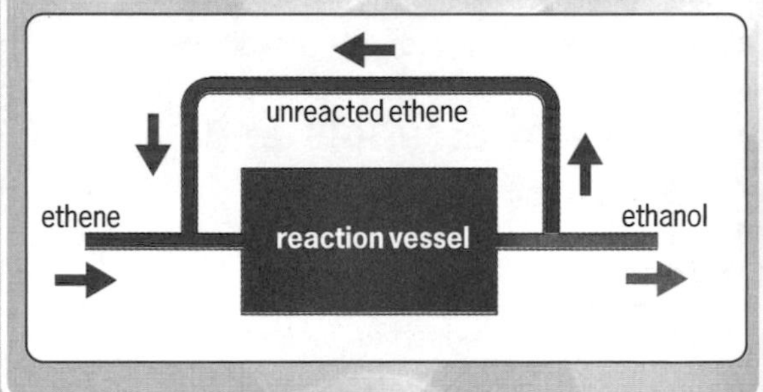

Ethanol is used as a solvent, as a fuel and is present in alcoholic drinks. There are two main methods for the production of ethanol.

Comparing the two methods

Method	Advantages	Disadvantages
Fermentation	• Uses renewable resources, e.g. corn and sugar cane	• Large amounts of raw materials and equipment are needed – making the process expensive
	• Good way to use waste organic material	• Impure ethanol is produced
		• Fractional distillation uses energy, which is expensive
		• Fermentation is slow
		• Because it is a batch process there are periods of time where no ethanol is being produced
		• At a certain concentration of ethanol the yeast cells become inactive, limiting the amount of ethanol produced per batch
From ethene	• Reaction is fast	• Ethene is obtained from crude oil, which is a finite resource
	• Plant runs continuously, so no time is wasted stopping and starting	• Energy is needed to make the steam, and to achieve the correct temperature and pressure for the reaction to occur
	• Reaction vessels are smaller	• Some ethene doesn't react, so it needs to be continuously recycled to give a satisfactory yield of ethanol
	• Process produces pure ethanol	

THE REACTIVITY SERIES AND DISPLACEMENT REACTIONS

The reactivity series

The reactivity series is a list of the most common metals placed in order of reactivity. Carbon and hydrogen are included for reference.

Potassium	K	Most reactive
Sodium	Na	
Calcium	Ca	
Magnesium	Mg	
Aluminium	Al	
Carbon	C	
Zinc	Zn	
Iron	Fe	
Lead	Pb	
Hydrogen	H	
Copper	Cu	
Silver	Ag	
Gold	Au	Least reactive

If you know about the position of metals in the reactivity series then you can predict how the metal will react.

- Metals above hydrogen will react with acids and with water.
- The higher the metal is above hydrogen then the more vigorous the reaction will be with water or acid.

Displacement reactions

A displacement reaction is one where **a more reactive substance takes the place of a less reactive substance** in a compound.

For example, if zinc is added to copper(II) sulphate solution then the solution slowly turns from blue to colourless and a red/brown solid is observed.

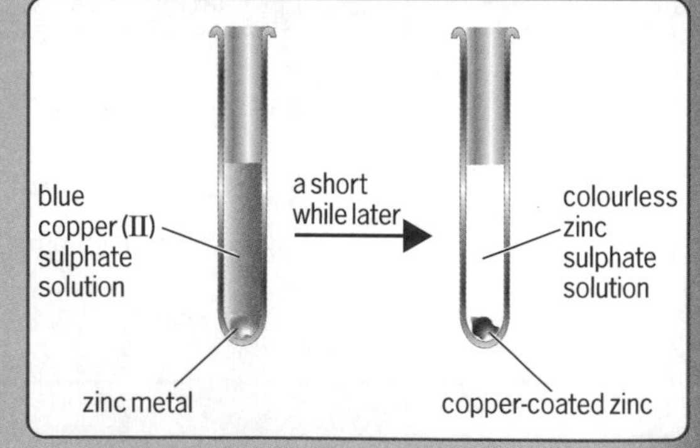

zinc + copper(II) sulphate → zinc sulphate + copper

In this reaction zinc **displaces** copper because zinc is more reactive than copper.

Understanding the reactivity series allows you to predict some chemical reactions and understand how metals are extracted from their ores.

Working out a reactivity series

The following experiments were set up. Small pieces of copper, zinc, iron and an unknown metal, X, were added to solutions of their sulphates in water. The results are shown in the table. A tick (✓) indicates that there was a reaction. A cross (×) indicates no reaction.

	Copper(II) sulphate	Zinc sulphate	Iron sulphate	Sulphate of metal X
Copper		×	×	×
Zinc	✓		✓	×
Iron	✓	×		×
Metal X	✓	✓	✓	

Metal X is the most reactive metal because it displaces all of the other metals from their sulphates.

Metal X could be magnesium. In this case, a metal such as sodium (which is more reactive than copper, zinc and iron) is not a sensible suggestion because sodium is such a reactive metal that it will react with the water.

Progress check

1. Name a metal that is more reactive than calcium.
2. Name a metal that is less reactive than lead.
3. Why would you not expect a piece of copper to react with hydrochloric acid?
4. Explain why gold does not readily corrode.
5. What is a displacement reaction?
6. Write an equation for the reaction that occurs when a piece of magnesium is added to a solution of zinc sulphate.
7. Explain why zinc will not react with magnesium sulphate.

IRON AND STEEL

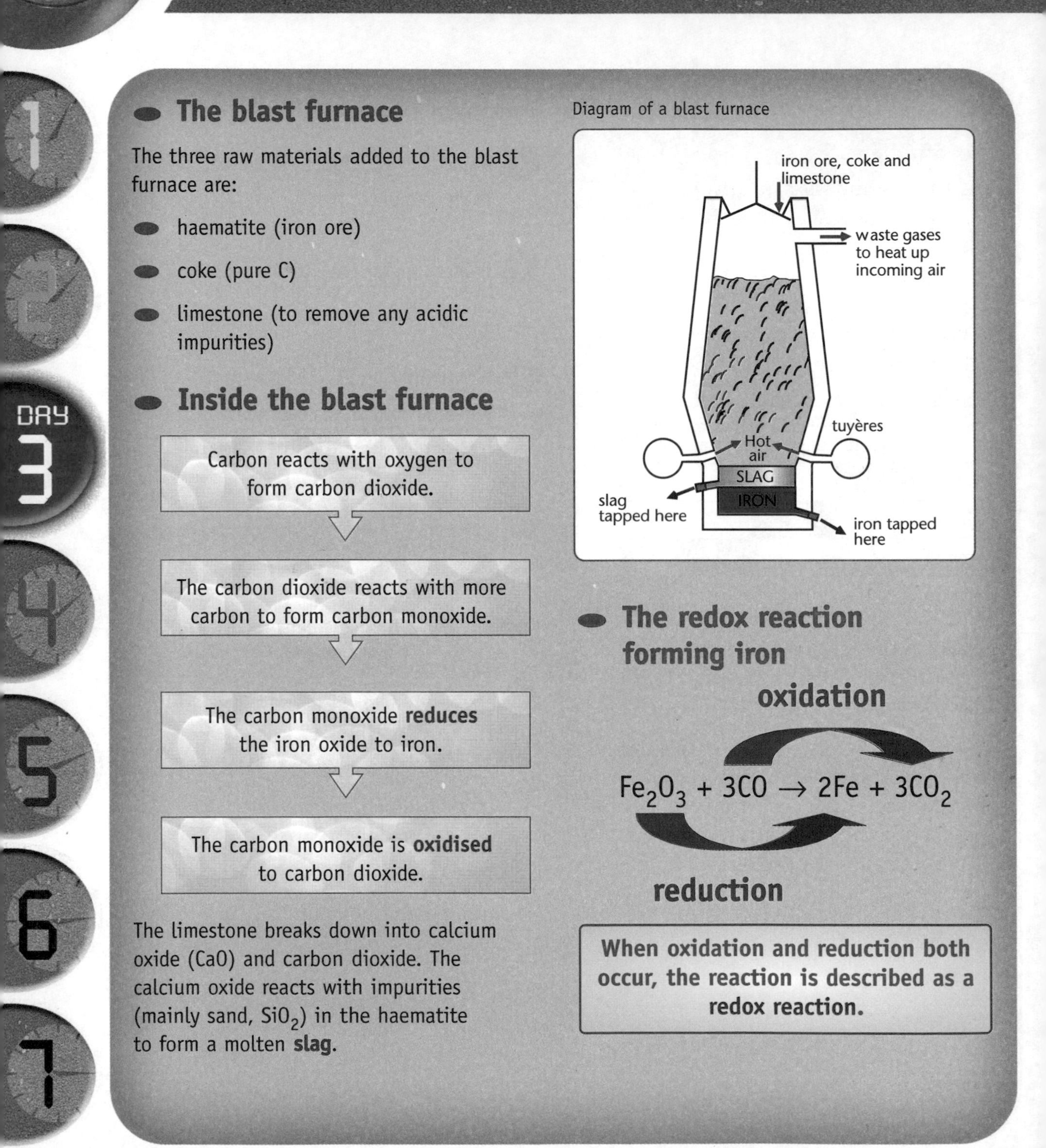

The blast furnace

The three raw materials added to the blast furnace are:

- haematite (iron ore)
- coke (pure C)
- limestone (to remove any acidic impurities)

Inside the blast furnace

Carbon reacts with oxygen to form carbon dioxide.

↓

The carbon dioxide reacts with more carbon to form carbon monoxide.

↓

The carbon monoxide **reduces** the iron oxide to iron.

↓

The carbon monoxide is **oxidised** to carbon dioxide.

The limestone breaks down into calcium oxide (CaO) and carbon dioxide. The calcium oxide reacts with impurities (mainly sand, SiO_2) in the haematite to form a molten **slag**.

Diagram of a blast furnace

The redox reaction forming iron

$$Fe_2O_3 + 3CO \rightarrow 2Fe + 3CO_2$$

When oxidation and reduction both occur, the reaction is described as a redox reaction.

Iron is extracted from its ore, haematite (Fe_2O_3), in a blast furnace. Most of the iron produced by this method is converted to steel.

The conversion of iron to steel

- Iron from the blast furnace has a lot of carbon mixed in with it. This makes the iron **brittle**.
- **Steel** is iron with less carbon in it.
- Other metals can also be added, forming **alloys**.
- The carbon is removed from the blast furnace iron by forcing oxygen gas through the molten iron.
- The oxygen will react with any impurities in the iron.

Types of steel

Stainless steel	Manganese steel
• Does not rust	• Very strong
• Contains 70% Fe 20% Cr 10% Ni	• Contains 85% Fe 13.8% Mn 1.2% C
• Used for cutlery	• Used for railway track

Progress check

1. Name the raw materials in the blast furnace.
2. What is the purpose of the limestone?
3. Label this diagram of the blast furnace to show where the iron is removed from the blast furnace.

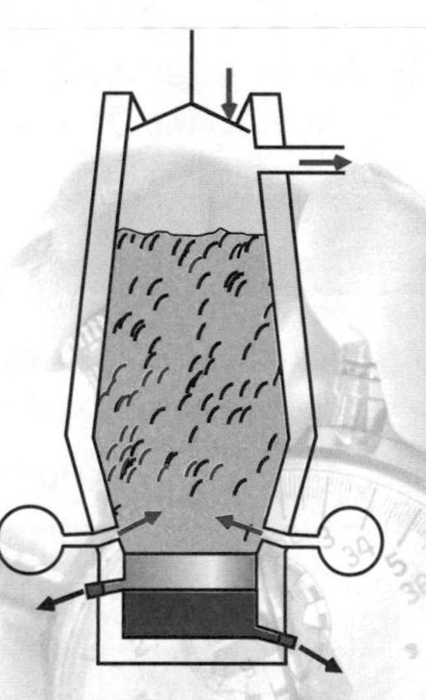

4. Which substance reduces the iron oxide?
5. How is iron converted to steel?
6. How does adding chromium affect the properties of steel?
   a) Reduces its brittleness
   b) Makes it a nicer colour
   c) Prevents it from rusting
   d) Increases its electrical conductivity

EXTRACTION OF ALUMINIUM AND ANODISING

Extraction of aluminium

Reactive metals such as aluminium are extracted by electrolysis.

The raw material for producing aluminium is purified **bauxite** (aluminium oxide, Al_2O_3). Aluminium oxide has a very high melting point. It is dissolved in **molten cryolite**, a compound that contains aluminium. Dissolving the purified bauxite in molten cryolite **lowers the melting point** of the aluminium oxide. This means that less heat is needed to keep the mixture molten and so energy costs are reduced.

The electrodes are made of graphite. The aluminium ions (Al^{3+}) are attracted to the negative electrode (**cathode**), where they gain electrons to form aluminium.

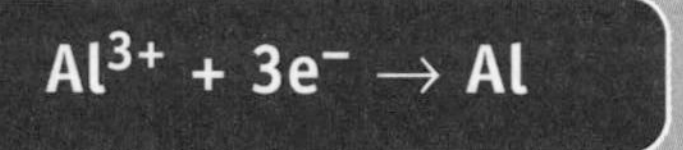

This is a **reduction** reaction. Reduction occurs when electrons are gained.

The oxide ions (O^{2-}) are attracted to the positive electrode (**anode**) where they lose electrons to form oxygen gas.

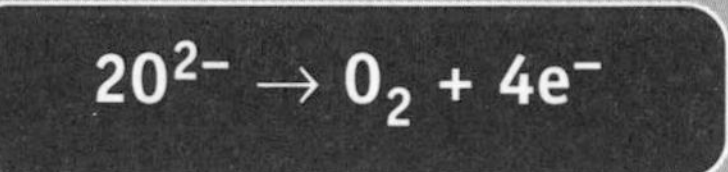

This is an **oxidation** reaction. Oxidation occurs when electrons are lost.

The oxygen formed at the anode reacts with the graphite to form carbon dioxide. This means that the anodes burn away and so need to be replaced regularly.

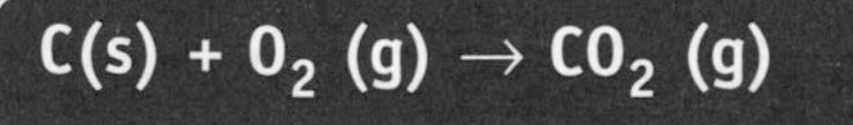

Oxidation	**R**eduction
Is	**I**s
Loss (of electrons)	**G**ain (of electrons)

Diagram of cell used for the industrial extraction of aluminium

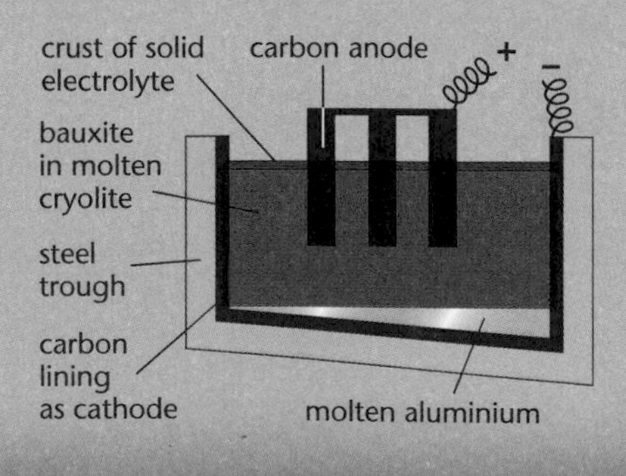

Aluminium is a more reactive metal than carbon. It is extracted from its ore by electrolysis.

Aluminium

Aluminium does not oxidise (corrode) as quickly as its position in the reactivity series would suggest. Exposure of aluminium to the atmosphere causes the aluminium to react with oxygen in the air, forming a thin **coating of aluminium oxide** on the surface. This acts as a **barrier** to water and oxygen and so prevents any further corrosion.

Aluminium is a useful structural metal. It can be made harder and stronger by mixing in small amounts of other metals to make **alloys**.

Anodising

The oxide coating on the aluminium can be artifically thickened by anodising. This involves making the aluminium the anode and using dilute sulphuric acid as the electrolyte. Oxygen gas forms at the anode and reacts with the aluminium.

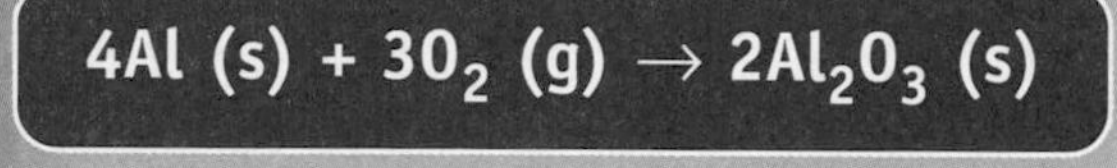

$$4Al\ (s) + 3O_2\ (g) \rightarrow 2Al_2O_3\ (s)$$

Progress check

1. Why does aluminium need to be extracted from its ore by electrolysis?
2. What is the name of the ore from which aluminium is extracted?
3. Complete the following half-equation showing what happens at the cathode.

   Al^{3+} + ___e^- → Al
4. Is this an oxidation or a reduction?
5. Explain your answer.
6. In the electrolysis of bauxite, why does the anode need to be regularly replaced?
7. Why does aluminium not react as rapidly with water as would be predicted from its position in the reactivity series?
8. Explain why aluminium is anodised.
9. Write a word equation to show what happens to the aluminium when it is anodised.

DAY 3

COPPER PURIFICATION AND TITANIUM PRODUCTION

Purification of copper

Copper can be purified by electrolysis, using a positive electrode made of the impure copper and a negative electrode of pure copper in a solution containing copper ions.

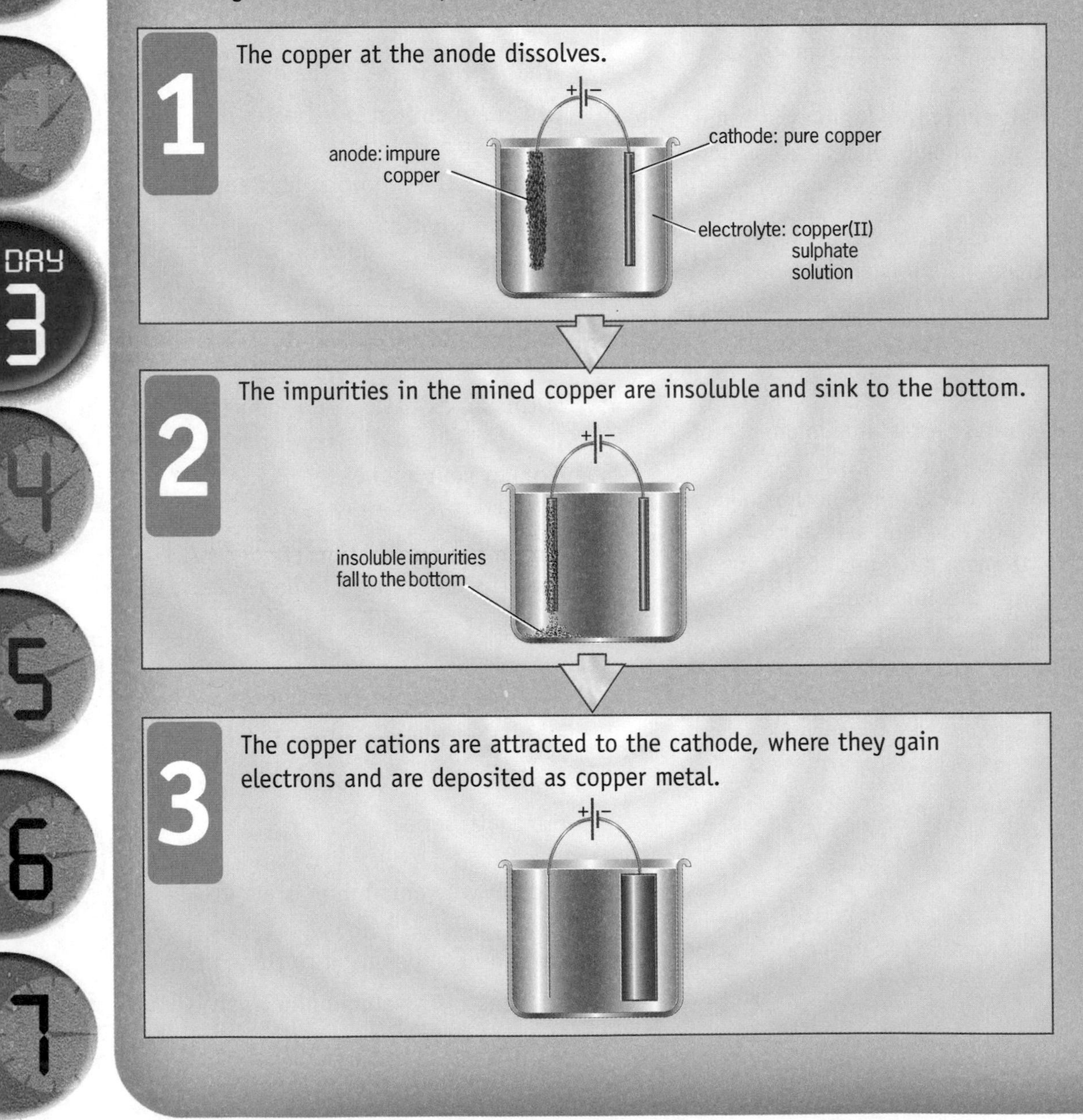

Very pure copper is a very good conductor of electricity. Mined impure copper is purified by electrolysis. Titanium is widely used in machinery due to its high tensile strength and low density.

Equations

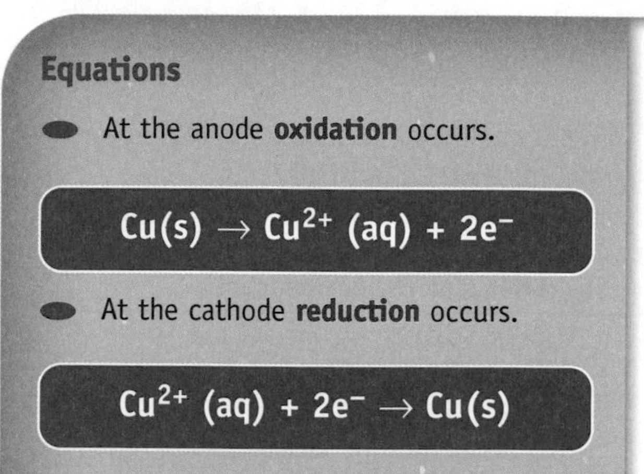

- At the anode **oxidation** occurs.

$$Cu(s) \rightarrow Cu^{2+}(aq) + 2e^-$$

- At the cathode **reduction** occurs.

$$Cu^{2+}(aq) + 2e^- \rightarrow Cu(s)$$

Titanium production

If titanium were to be placed in the reactivity series, it would be located below carbon. This would suggest that titanium should therefore be extracted from its ore by heating with carbon (i.e. in a blast furnace). However, due to economic reasons a different method is used.

The ore rutile (TiO_2) is first converted into titanium chloride ($TiCl_4$). This is then reduced by heating with sodium or magnesium.

titanium(IV) chloride + sodium
→ titanium + sodium chloride

$$TiCl_4 + 4Na \rightarrow Ti + 4NaCl$$

Titanium is extracted in the absence of air because the oxygen in the air would react with the sodium metal.

Progress check

1. Why does copper need to be purified?

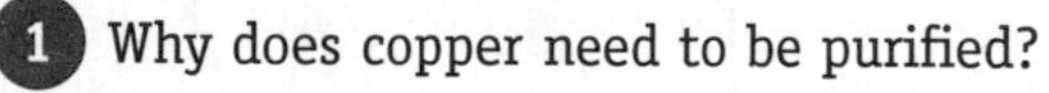

2. Label this diagram to show how an impure piece of copper can be purified.

3. Balance the following half-equation to show what happens at the cathode during the purification of copper.

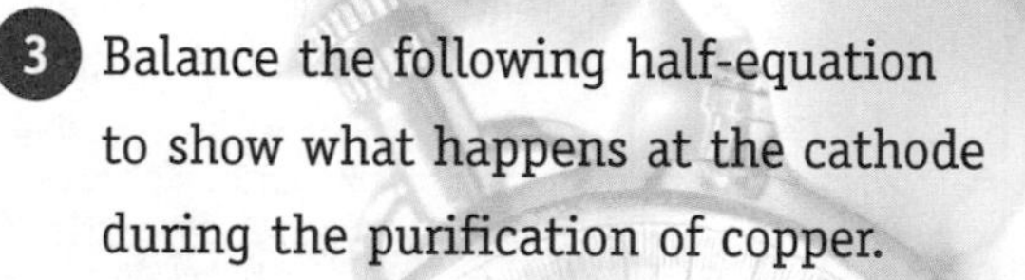

Cu^{2+} + ___$e^- \rightarrow Cu$

4. Titanium can be obtained from titanium(IV) chloride by reaction with magnesium. Write a word equation to represent this process.

5. When titanium is extracted from titanium(IV) chloride, air is excluded from the reaction vessel.
Explain why.

DAY 3

POLLUTION AND USEFUL PRODUCTS FROM LIMESTONE

Pollution

Most chemical industrial processes require a lot of energy. Most of this energy comes from the burning of **fossil fuels**: coal, gas, oil and petrol. The burning of fossil fuels produces many **pollutants**. These pollutants and their effects on the environment are shown below.

Pollutant	How is it formed?	Effects on the environment
Carbon dioxide	Burning of fossil fuels (all of which contain carbon)	Carbon dioxide is a greenhouse gas This contributes to global warming
Carbon monoxide	Incomplete combustion of fossil fuels	It is a poisonous gas It reacts with the haemoglobin in red blood cells and stops it carrying oxygen to the cells
Sulphur dioxide	Many fossil fuels contain traces of sulphur in them. When sulphur is burnt sulphur dioxide is formed	Sulphur dioxide is not good for people with respiratory problems Sulphur dioxide also causes acid rain
Nitrogen oxides	Extreme heat (and sparks inside engines) cause oxygen and nitrogen in the air to react with each other	As sulphur dioxide

Formation of limestone

Limestone is made from the shells and skeletons of dead sea creatures, which accumulated as a sediment on the sea floor.

Heat and pressure over millions of years turned this sediment into rock.

Limestone

Limestone, which is mainly calcium carbonate ($CaCO_3$), can be quarried and used as a building material. Powdered limestone can be used to neutralise acidity in lakes and soils.

When limestone is heated in a kiln it breaks down into quicklime (calcium oxide, CaO) and carbon dioxide. This is a **thermal decomposition** reaction. Other metal carbonates behave in a similar way.

$$CaCO_3 \rightarrow CaO + CO_2$$

calcium carbonate (limestone) → calcium oxide (quicklime) + carbon dioxide

Limestone is one form of calcium carbonate ($CaCO_3$). It is used to make cement and glass. Most chemical industries produce a lot of pollution.

Quicklime reacts with water to produce slaked lime (calcium hydroxide, $Ca(OH)_2$), which is also used to reduce the acidity of soil.

$$CaO + H_2O \rightarrow Ca(OH)_2$$

calcium oxide (quicklime) + water → calcium hydroxide (slaked lime)

Cement

Cement is produced by roasting powdered limestone with powdered clay in a rotary kiln.

When cement is mixed with water, sand and crushed rock, a slow chemical reaction produces a hard, stone-like building material called **concrete**.

Glass

Glass is made by heating a mixture of limestone, sand and soda (sodium carbonate).

Progress check

1. Fill in the missing substances in the reactions of limestone.

   $CaCO_3 \rightarrow CaO +$ ______

   $CaO +$ ______ $\rightarrow Ca(OH)_2$

2. To make concrete, powdered limestone and powdered clay are roasted together, and mixed with ________, ________ and ________.

3. Glass is made by reacting limestone, soda (sodium carbonate) and ________.

4. For a named pollutant, describe:
   i. its formation
   ii. the effect it has on the environment

REVERSIBLE REACTIONS AND EQUILIBRIUM

Reversible reactions

If A and B react together to make C and D, and C and D react together to make A and B, this is a reversible reaction and it can be shown like this.

$$A + B \rightleftharpoons C + D$$

ammonium chloride ⇌ ammonia + hydrogen chloride

white solid ⇌ colourless gases

Equilibrium

When a reversible reaction occurs in a closed system (i.e. a sealed container), an equilibrium is reached when **the forward reaction occurs at exactly the same rate as the reverse reaction**.

The relative amounts of all the reacting substances at equilibrium depend on the conditions of the reaction. If the conditions are changed on a reaction at equilibrium then the reaction changes. The table below outlines the effect on the yield (i.e. the forward reaction) if temperature changes are made on reactions at equilibrium. The effect on the yield depends on whether the reaction is **exothermic** or **endothermic**.

Reaction	Change in temperature	Effect on yield (i.e. amount of product)
Exothermic	Increased	Decreased
	Decreased	Increased
Endothermic	Increased	Increased
	Decreased	Decreased

In gaseous reactions, an increase in pressure will favour the reaction which produces the least number of molecules, as shown by the symbol equation for that reaction.

Example

In the formation of ammonia:

$$N_2 + 3H_2 \rightleftharpoons 2NH_3$$

a high pressure will favour the forward reaction, i.e. the formation of ammonia, because there are two molecules of gas on the right-hand side of the equation and four on the left.

These factors, together with reaction rates, are important when determining the optimum conditions in industrial processes, including the Haber process (see page 36).

Reversible reactions go forwards and backwards. When the rate of the forward reaction is the same as the rate of the reverse reaction then the reaction is at equilibrium.

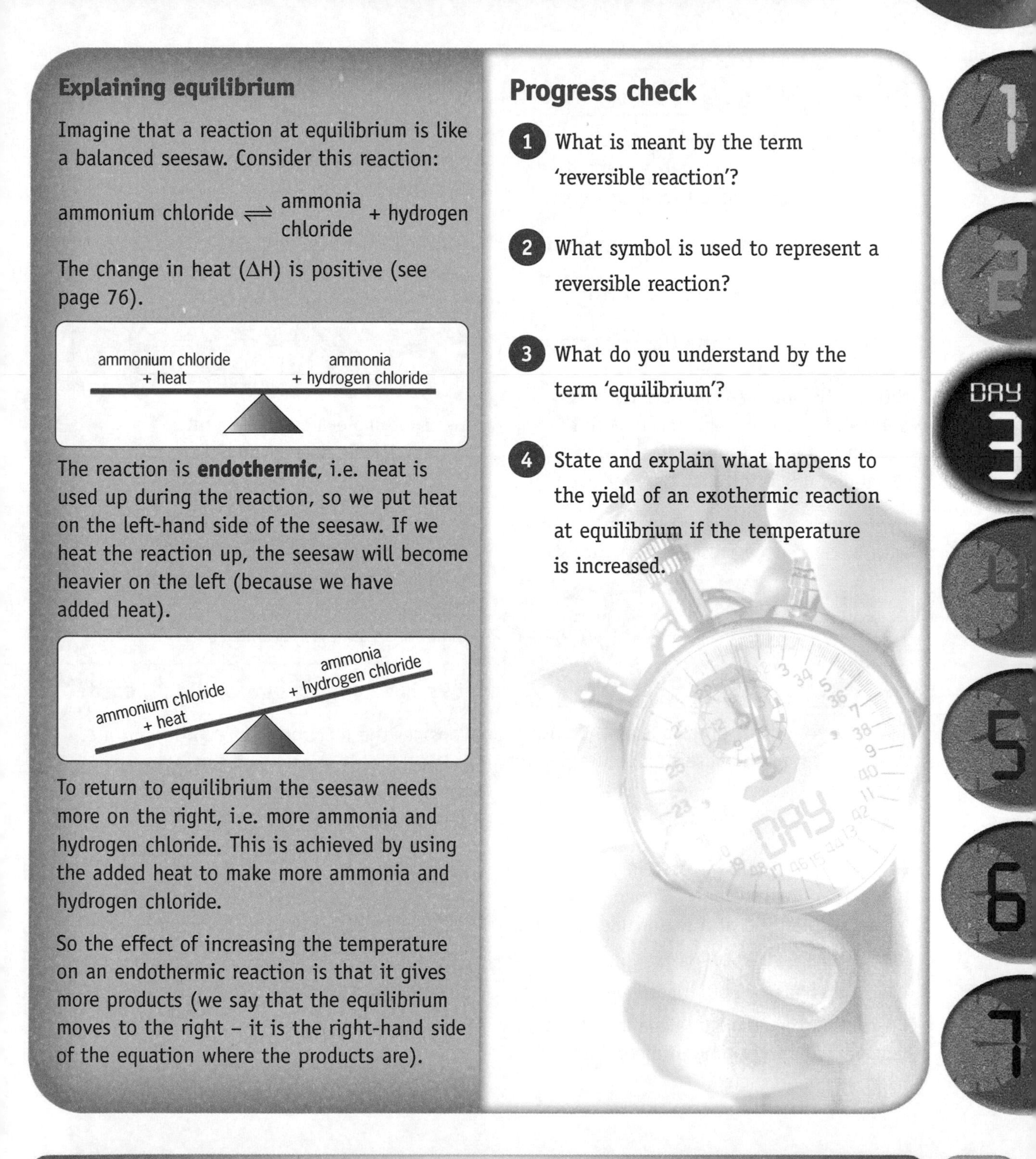

Explaining equilibrium

Imagine that a reaction at equilibrium is like a balanced seesaw. Consider this reaction:

ammonium chloride $\rightleftharpoons$ ammonia chloride + hydrogen

The change in heat (ΔH) is positive (see page 76).

The reaction is **endothermic**, i.e. heat is used up during the reaction, so we put heat on the left-hand side of the seesaw. If we heat the reaction up, the seesaw will become heavier on the left (because we have added heat).

To return to equilibrium the seesaw needs more on the right, i.e. more ammonia and hydrogen chloride. This is achieved by using the added heat to make more ammonia and hydrogen chloride.

So the effect of increasing the temperature on an endothermic reaction is that it gives more products (we say that the equilibrium moves to the right – it is the right-hand side of the equation where the products are).

Progress check

1. What is meant by the term 'reversible reaction'?
2. What symbol is used to represent a reversible reaction?
3. What do you understand by the term 'equilibrium'?
4. State and explain what happens to the yield of an exothermic reaction at equilibrium if the temperature is increased.

MANUFACTURE OF AMMONIA

Raw materials

Ammonia is made by reacting nitrogen and hydrogen together in the presence of an iron catalyst.

nitrogen + hydrogen ⇌ ammonia

$$N_2 + 3H_2 \rightleftharpoons 2NH_3$$

This is known as the Haber process.

Nitrogen is obtained from air. Hydrogen is burnt in the air and this reacts with the oxygen in the air forming liquid water. The remaining gas is virtually pure nitrogen.

Hydrogen is obtained from reacting methane with water (in the presence of a **catalyst**).

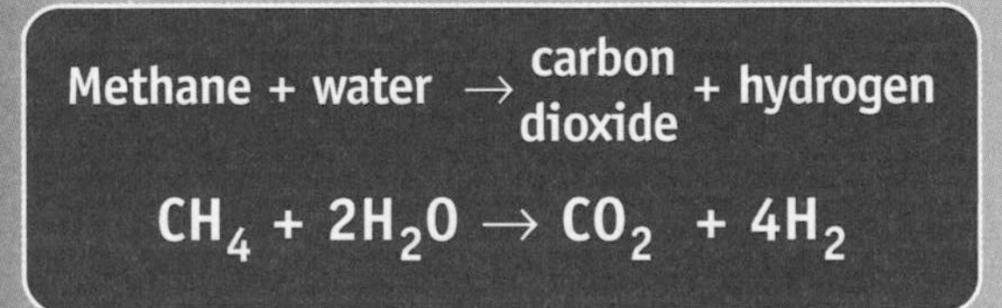

The catalyst in this process is iron, thinly coated onto metal gauze to give it a greater surface area.

Conditions

The reaction conditions used in the manufacture of ammonia, chosen to produce a worthwhile yield at a reasonable rate, are:

- a temperature of 450°C
- a pressure of 200 atmospheres

On cooling the ammonia liquefies and is removed. The remaining nitrogen and hydrogen are recycled.

Why is a pressure of 200 atmospheres used?

A high pressure favours the production of ammonia because **high pressure favours the reaction that produces fewer molecules of gas.** There are four molecules of reactant gases (one nitrogen and three hydrogen) but only two molecules of product.

$$1N_2(g) + 3H_2(g) \rightleftharpoons 2NH_3(g)$$

1 + 3 = 4 molecules of reactant
2 molecules of product

An even higher pressure would produce more ammonia but achieving very high pressures costs more than the value of the extra ammonia produced. So a pressure of 200 atmospheres is a compromise between yield (amount of ammonia) and cost.

Why is a temperature of 450°C used?

Consider the reaction is at equilibrium, i.e. balanced. This means that the rate at which nitrogen and hydrogen react to make ammonia is the same as the rate at which ammonia breaks down to form nitrogen and hydrogen.

Remember, this is an exothermic reaction, so heat is a product.

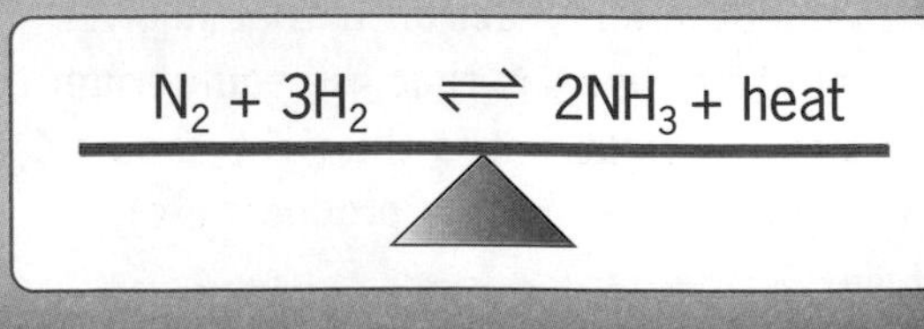

Ammonia (NH_3) is a very useful chemical. It is used in the manufacture of fertilisers, explosives and nitric acid. Ammonia is made using the Haber process.

If the temperature is increased (i.e. more heat added) then the reaction will no longer be at equilibrium.

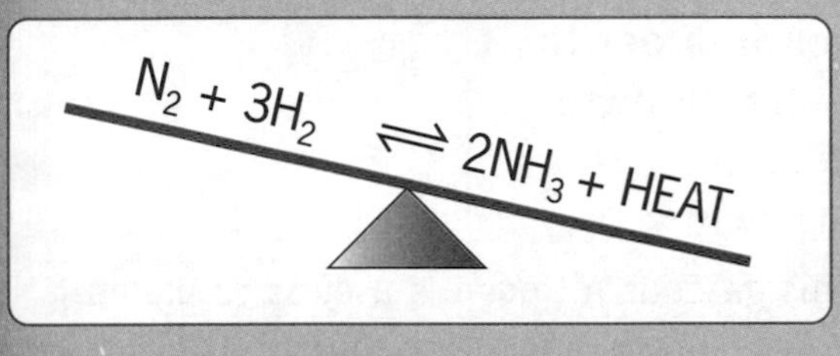

In order to restore equilibrium, more nitrogen and hydrogen need to be formed.

This is achieved by ammonia breaking down to form nitrogen and hydrogen, i.e. the rate of the reverse reaction will become greater than the rate of the forward reaction until the balance (the equilibrium) has been restored. So a high temperature causes more ammonia to break down. The higher the temperature then the more ammonia will break down.

So, if you want to make a lot of ammonia then you need a low temperature. However, if the temperature is too low then the rate of the reaction will be too slow!

So a temperature of 450°C is a compromise between yield of ammonia and rate of reaction.

Progress check

1. What temperature and pressure are used in the production of ammonia?
2. Explain why these conditions are used.
3. Give one use for ammonia.
4. What happens to the yield of ammonia if the temperature is increased?
5. What happens to the yield of ammonia if the pressure is increased?
6. Why is a temperature of less than 450°C not used?

USES OF AMMONIA

Fertilisers

A fertiliser is a substance that promotes plant growth.

Plants need nitrogen in order to make proteins for plant growth, repair of cells etc.

However, whilst the air is almost 80% nitrogen, this gaseous nitrogen is useless to plants. They need nitrogen in a soluble form so that they can absorb it in water through their roots.

Nitrogen-based fertilisers contain nitrogen in a soluble form, usually as **nitrates**.

Fertilisers can be made by neutralising ammonia with an acid.

- If nitric acid is used then the fertiliser produced is **ammonium nitrate**.

ammonia + nitric acid → ammonium nitrate

$$NH_3 + HNO_3 \rightarrow NH_4NO_3$$

- If sulphuric acid is used then the fertiliser produced is **ammonium sulphate**.

ammonia + sulphuric acid → ammonium sulphate

$$2NH_3 + H_2SO_4 \rightarrow (NH_4)_2SO_4$$

Fertilisers promote crop growth.

Ammonia is widely used to make many useful substances. These include fertilisers and nitric acid. Overuse of fertilisers can cause environmental problems.

10 MINS

Overuse of fertilisers

Rainwater can dissolve fertilisers and carry them into streams, rivers and ponds etc. This process is called **leaching**.

A build-up of fertiliser in water can cause an excessive growth of algae (known as an **algal bloom**). This can prevent sunlight reaching other plants, which then die and start to decay. This decaying process uses up oxygen, which eventually leads to the death of fish etc. Then these too will start to decay. Eventually there will be no life left in the river, at which point the river is said to be dead.

This whole cycle is known as **eutrophication**.

leaching → algal bloom → death and decay of algae → lack of oxygen → death of remaining plants and animals

Why do chemists prefer *nitrates?*

Because they're cheaper than *day-rates!*

There is evidence that too many nitrates in the domestic water supply can cause health problems, e.g. blue-baby syndrome.

Nitric acid

Nitric acid (HNO_3) is used to make fertilisers, explosives and drugs. The raw materials in the manufacture of nitric acid are ammonia, air and water.

The diagram shows the stages in the manufacture of nitric acid.

It is important that the gases are not allowed to escape into the atmosphere because they cause acid rain.

Ammonia is reacted with oxygen in the presence of a platinum/rhodium catalyst to form nitrogen monoxide.
$4NH_3 + 5O_2 \rightarrow 4NO + 6H_2O$

↓

The nitrogen monoxide is cooled and more air added. The nitrogen monoxide reacts to form nitrogen dioxide.
$2NO + O_2 \rightarrow 2NO_2$

↓

The nitrogen dioxide reacts with air and water to form nitric acid.
$4NO_2 + O_2 + 2H_2O \rightarrow 4HNO_3$

WORKING OUT FORMULAE

Formulae

The formula of a substance simply tells you how many of each type of atom there is in a compound. The formula of water is H_2O. This means that in a molecule of water there are two atoms of hydrogen and one atom of oxygen.

To work out the formula of a compound you need to know about the electron configuration of each atom present.

Water contains hydrogen atoms and oxygen atoms.

Hydrogen has the electron configuration: 1

Remember, all atoms want to have a full outer shell of electrons so an atom of hydrogen wants to gain one electron. We say that hydrogen has a **valency** of 1.

Oxygen has the electron configuration: 2,6

Oxygen wants to gain two electrons so oxygen has a valency of 2.

To work out the formula of water simply write the atoms showing their valencies like this:

$H^1 \ O^2$

Then swap the valencies like this ...

$H^1 \times O^2$

If you follow the arrows then this gives you the formula H_2O_1. We do not usually write the '1' so the formula simply becomes H_2O.

Example

Work out the formula of aluminium sulphide.

The electron configuration of Al is 2,8,3 so an atom of aluminium wants to lose three electrons, i.e. its valency is 3.

The electron configuration of sulphur is 2,8,6 so an atom of sulphur wants to gain two electrons, i.e. its valency is 2.

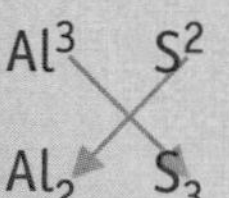

Hence, the formula of aluminium sulphide is Al_2S_3.

Formulae involving complex ions

The above examples were for compounds containing two elements. Complex ions contain two or more atoms. The table below shows the formulae of the common complex ions.

Formula of ion	Valency	Name of ion
OH^-	1	Hydroxide
NO_3^-	1	Nitrate
SO_4^{2-}	2	Sulphate
CO_3^{2-}	2	Carbonate
NH_4^+	1	Ammonium

An important skill at GCSE is the ability to be able to write down the formula of a substance. With a small amount of practice this can easily be mastered.

With complex ions it is important to remember that the charge on the ion applies to the whole ion.

The formula can be worked out as it was before but because the charge applies to the whole ion, put brackets around the atoms of the complex ion before you swap the valencies.

Example

The formula of calcium hydroxide

$Ca^{2}\ (OH)^{1}$

$Ca_{1}\ (OH)_{2}$

Hence, the formula is $Ca(OH)_2$.

Example

The formula of ammonium sulphate

$(NH_4)^{1}\ (SO_4)^{2}$

$(NH_4)_2\ (SO_4)_1$

In the above formula you can remove the brackets from around the SO_4 because the number there is 1. Hence, the formula of ammonium sulphate becomes:

$(NH_4)_2SO_4$

Progress check

Work out the formula of each of these compounds:

1. sodium oxide
2. magnesium chloride
3. carbon sulphide
4. sodium carbonate
5. lithium nitrate
6. aluminium hydroxide

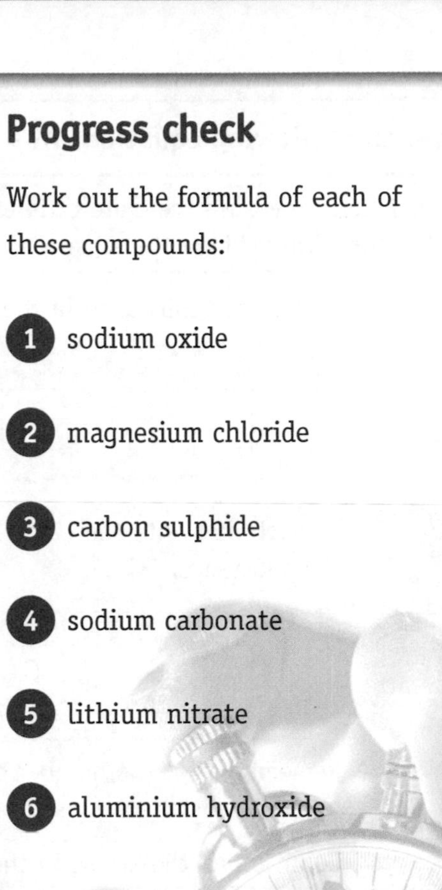

BALANCING EQUATIONS

Consider the following equation: $Al + Cl_2 \rightarrow AlCl_3$

It does not balance because there are different numbers of each type of atom on either side of the equation.

STEP 1 Under each substance in the equation draw a box.

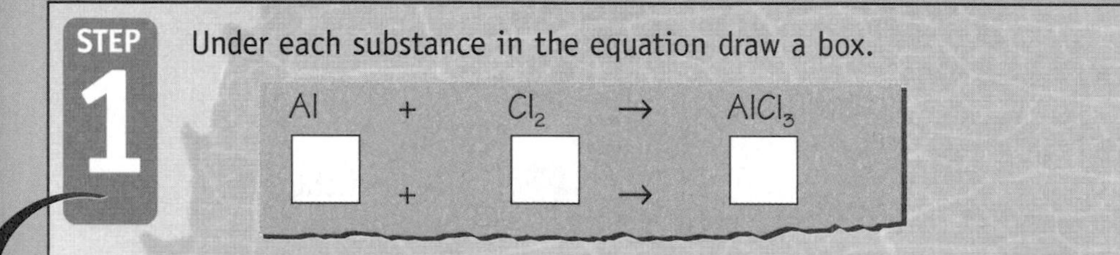

STEP 2 In each box write the correct numbers of atoms of each element in the substance.

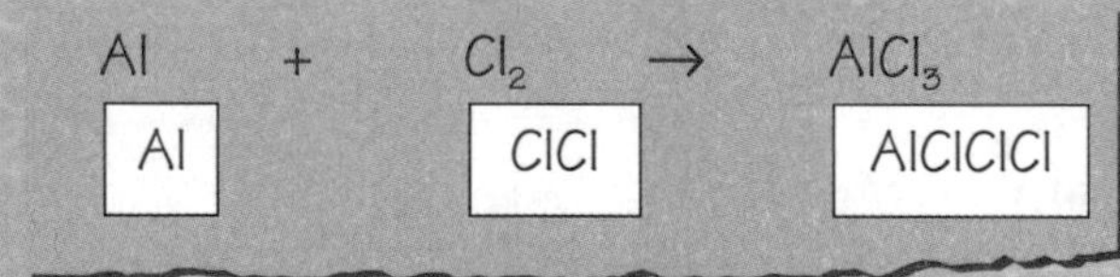

Remember that the number by an element tells you how many atoms of that element are present.

STEP 3 Count up all the atoms in the boxes! There must be the same number of atoms of each element on both sides of the equation. You can add as many boxes as you like but you are not allowed to put things in or take things out of a box.

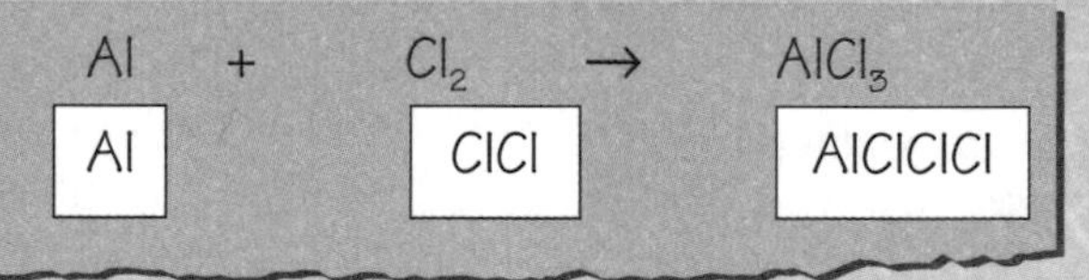

The numbers of Al atoms already balance but there are three Cl atoms on the right but only two on the left, so add another box of Cl_2 to the left!

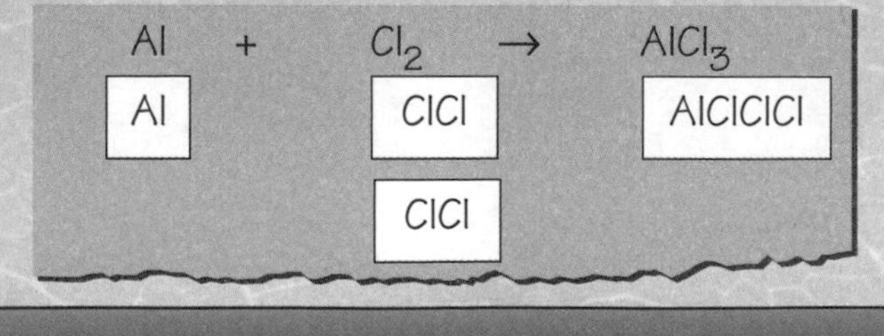

Many students think that balancing equations is impossible! To overcome this, use the technique on these pages for balancing equations.

10 MINS

There are now four Cl atoms on the left but only three on the right, so add another box of $AlCl_3$ to the right.

Al	+	Cl_2	→	$AlCl_3$
Al		ClCl		AlClClCl
		ClCl		AlClClCl

Now there are six Cl atoms on the right but only four on the left so add another box of Cl_2 to the left.

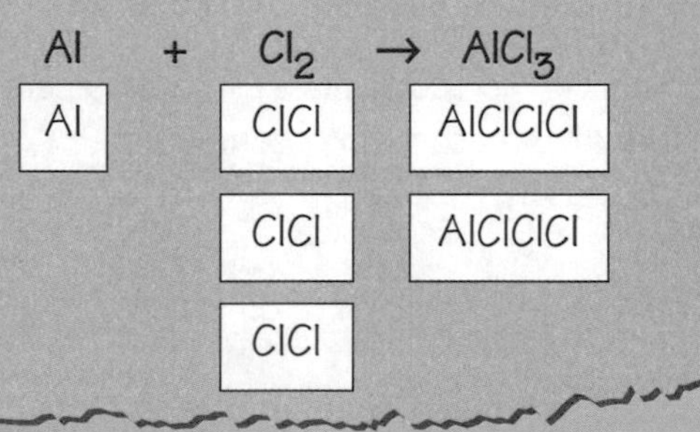

Now the Cl balances. However, there are now two Al atoms on the right but only one on the left. Add one more box of Al on the left.

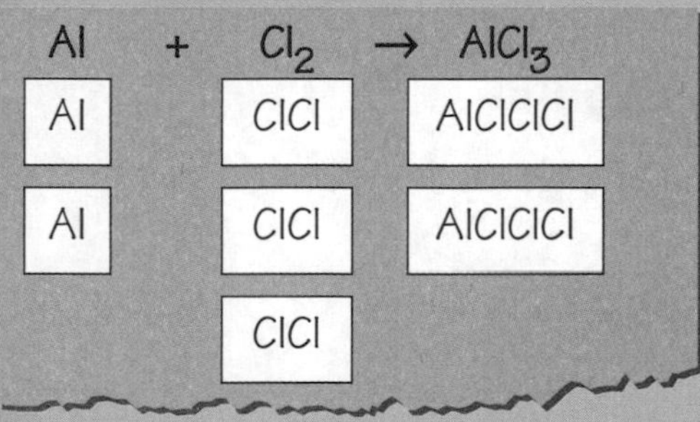

The equation now balances! To write the balanced equation you simply count up the number of boxes.

$$2Al + 3Cl_2 \rightarrow 2AlCl_3$$

Progress check

Balance these equations.

1. $Ca + O_2 \rightarrow CaO$
2. $Cr + HCl \rightarrow CrCl_3 + H_2$
3. $CH_4 + Cl_2 \rightarrow CCl_4 + HCl$
4. $Na + O_2 \rightarrow Na_2O$
5. $P + Br_2 \rightarrow PBr_3$
6. $Al + O_2 \rightarrow Al_2O_3$

MOLE CALCULATIONS 1

Mass of atoms and molecules

Atoms of different elements have different masses. To be able to work out exactly what is happening in chemical reactions you need to know how the masses of atoms compare with each other, i.e. their **relative atomic masses** (A_r).

> **The relative atomic mass of an atom is the mass of an atom compared to the mass of an atom of carbon.**

When you want to know the relative atomic mass of an atom you just use the **mass number** (large number) on the periodic table (you are provided with a periodic table in the exam).

The **relative formula mass** (M_r, sometimes called the relative molecular mass) of a molecule is found by adding together the individual atomic masses of the atoms in a compound.

Examples

For HF the M_r is:

$1 + 19 = 20$

For $CaCO_3$ the M_r is:

$40 + 12 + (3 \times 16) = 100$

Percentage of an element in a compound

The percentage of an element in a compound is found by working out:

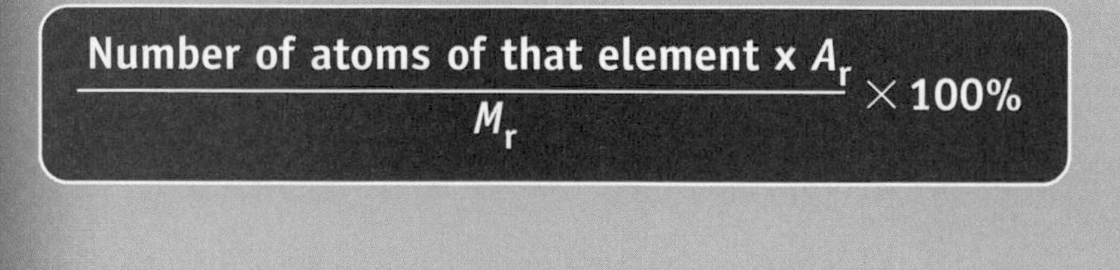

$$\frac{\text{Number of atoms of that element x } A_r}{M_r} \times 100\%$$

Many students find mole calculations difficult. Learn to follow a few simple steps and, with lots of practice, you will find them much easier!

10 MINS

Example

Calculate the percentage of calcium in calcium oxide (CaO).

$$\frac{1 \times 40}{56} \times 100\% = 71\%$$

Empirical formula calculations

The empirical formula of a compound is the simplest whole-number ratio of atoms in a compound.

You will be given the masses of elements that are present in a compound. You can find the empirical formula in two steps.

1 Divide each mass by the A_r (this is the number of moles).

2 Divide each of the above answers by the smallest answer in **1**.

Example

Calculate the empirical formula of a compound that contains 2.8 g of silicon and 3.2 g of oxygen.

	Si	**O**
Mass (g)	2.8	3.2
1) moles		
= mass ÷ A_r	2.8 ÷ 28	3.2 ÷ 16
=	0.1	0.2
2) ÷ 0.1	0.1 ÷ 0.1	0.2 ÷ 0.1
=	1	2

Hence, the empirical formula is SiO_2.

Progress check

1 Calculate the relative formula mass of:

i. MgO

ii. H_2SO_4

iii. $Na_2S_2O_3$

2 Using your answers to question 1, work out the percentage by mass of:

i. oxygen in MgO

ii. sulphur in H_2SO_4

iii. sodium in $Na_2S_2O_3$

3 Calculate the empirical formula of:

i. a compound containing 1.4 g of lithium and 1.6 g of oxygen

ii. a compound containing 0.28 g of silicon and 1.42 g of chlorine

MOLE CALCULATIONS 2

Calculating masses from equations

You can use the mass of either a product or a reactant in an equation to work out the mass of a product formed or reactant needed. To do this you also need a balanced symbol equation.

Follow these steps.

1 Work out M_r for each substance (this will be the same as A_r for elements). Write this under each species in the equation. If a substance has a number before it in the balanced equation then multiply A_r or M_r accordingly.

2 Under the appropriate substance write down the mass given in the question.

3 Work out the ratio between the value in step 1 and step 2 by dividing the value in step 1 by the value in step 2.

4 Divide the value for the substance you are interested in by the value obtained in step 3.

This is complicated! Use the worked examples that follow to apply the above steps.

Example

Calculate the mass of carbon dioxide formed if 6 g of carbon reacts with oxygen.

$C(s) + O_2(g) \rightarrow CO_2(g)$

	C	O_2	CO_2
Step 1	12	32	44
Step 2	6		
Step 3	12 ÷ 6 = 2		
Step 4			44 ÷ 2 = 22

Hence, the answer to the question is **22 g**.

Questions involving reacting masses are common at GCSE. Again, if you follow the steps and practise lots of questions then you will get much better at them!

Example

Calculate the mass of phosphorus trichloride (PCl_3) made if 3.1 g of phosphorus reacts.

$2P (s) + 3Cl_2 (g) \rightarrow 2PCl_3 (s)$

	$2P$	$3Cl_2$	$2PCl_3$
Step 1	62	213	275
Step 2	3.1		
Step 3	62 ÷ 3.1 = 20		
Step 4			275 ÷ 20 = 13.75

Answer = 13.75 g

Calculating volumes of gases from equations

This method can also be used to work out the volume of gases produced during a reaction. You need to know that the relative formula mass of any gas has a volume of 24 000 cm^3 (this is Avogadro's Law).

In the first example in the previous section you can work out the volume of carbon dioxide produced.

This is because M_r for carbon dioxide is 44 g. This means 44 g of CO_2 will occupy a volume of 24 000 cm^3.

The answer to the question was that 22 g of carbon dioxide was formed. Now, 22 is half of 44, therefore half of 24 000 cm^3 gas was formed, i.e. 12 000 cm^3.

Progress check

1. Calculate the mass of calcium chloride that can be made from reacting 5 g of calcium with chlorine.
   $Ca + Cl_2 \rightarrow CaCl_2$
2. Calculate the mass of sodium oxide that is made when 5.75 g of sodium is reacted with oxygen.
   $4Na + O_2 \rightarrow 2Na_2O$
3. Calculate the mass of aluminium that is needed to make 6 g of aluminium oxide.
   $4Al + 3O_2 \rightarrow 2Al_2O_3$
4. Calculate the volume of oxygen that is needed to make 8 g of magnesium oxide.
   $2Mg + O_2 \rightarrow 2MgO$

HOW TO USE THE QUIZ CARDS

There are several stages to successful revision – one of the most important is writing a list of the topics you need to know.

Then it's all about working through these essential topics, making useful notes and learning the key facts.

This is where these quiz cards can help you.

The questions on the cards provide a last-minute check of some key GCSE facts.

- You can leave them in the book and refer to them when you want
- You can tear them out and keep them handy for testing yourself

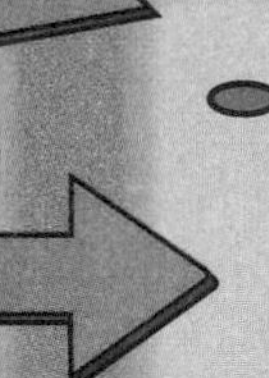

- You can get someone else to test you
- You can test your friends, which is also a good way of helping information sink in
- You can add to the cards by making your own sets of questions and answers

Remember – PREPARATION and PRACTICE and you'll be on the way to a good result!

Explain why ionic compounds do not conduct electricity when they are solid but do when molten or aqueous.

What two conditions are needed to crack long-chain alkanes?

Why do the alkali metals (group 1) become more reactive as you go down the group?

Draw the repeating unit formed from the polymerisation of this monomer.

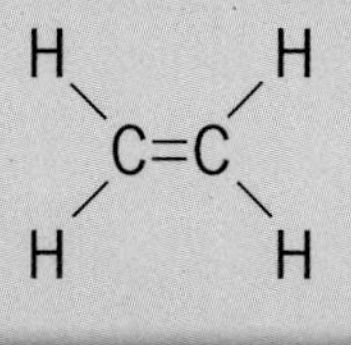

Label this diagram of a blast furnace.

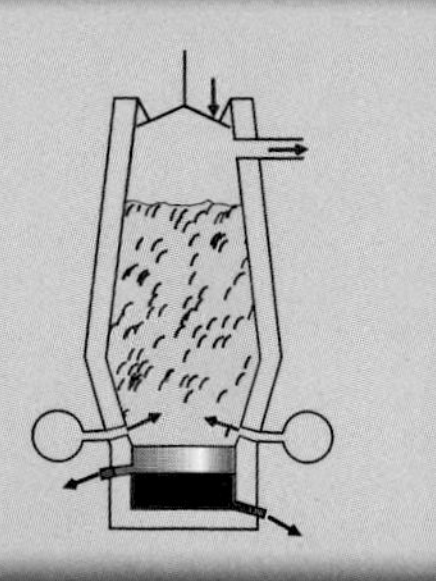

Write an ionic equation for a neutralisation reaction.

Write a word equation for the reaction between zinc and copper oxide.
Explain why the reaction occurs.

Balance this equation.

$Fe + O_2 \rightarrow Fe_2O_3$

Draw the structures of ethanol and ethanoic acid.

Why do substances with simple covalent structures have low melting points but those with giant covalent structures have high melting points?

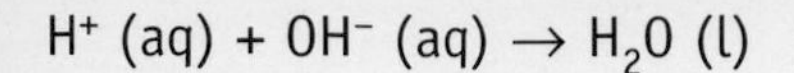

$H^+ (aq) + OH^- (aq) \rightarrow H_2O (l)$

In a solid the ions are not mobile (because they are only able to vibrate). When molten or aqueous the ions are free moving.

zinc + copper oxide → zinc oxide + copper

Zinc is more reactive than copper so it displaces the copper in copper oxide.

Heat and a catalyst.

$4Fe + 3O_2 \rightarrow 2Fe_2O_3$

As the atoms get larger the force of attraction between the nucleus and the outer shell electron decreases. Hence, less energy is needed to remove the electron and so reactivity increases.

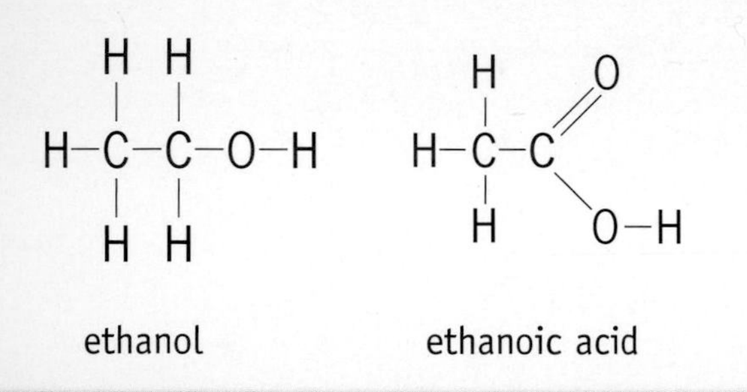

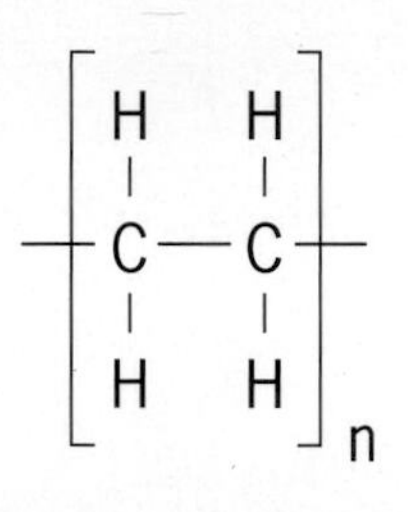

There are weak forces between the molecules in simple covalent structures which do not require much energy (heat) to overcome them. In giant covalent structures there are lots of strong covalent bonds requiring much more energy to break them.

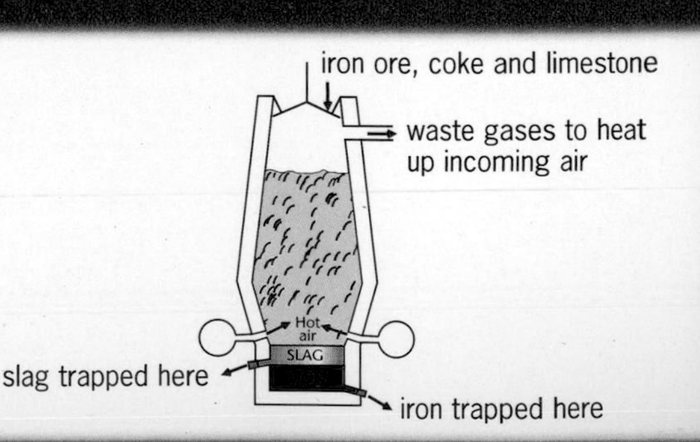

Why are the noble gases inert (chemically unreactive)?

Calculate the empirical formula of the substance that contains 1.1 g of boron and 10.65 g of chlorine.

What are isotopes?

What flame colour would be seen if a flame test was carried out on a compound containing potassium?

How is sedimentary rock converted into metamorphic rock?

Describe the test for the Cl^- ion. State what would be seen.

Draw the two isomers of the alkane with the molecular formula C_4H_{10}.

Calculate the mass of water produced when 6 g of hydrogen burns.

$2H_2 + O_2 \rightarrow 2H_2O$

What are the three products of the electrolysis of brine (concentrated sodium chloride solution)?

How can you show the presence of hardness in water?

BCl_3

Atoms of noble gases have full outer shells, meaning that they do not wish to gain, lose or share any electrons hence they do not react with other substances.

lilac

Two atoms (of the same element) that have the same number of protons but different numbers of neutrons.

Add a small amount of dilute nitric acid followed by silver nitrate solution. A white precipitate would be observed.

Movements in the Earth's surface cause the sedimentary rock to be exposed to high pressures and high temperatures which changes sedimentary rock into metamorphic rock.

54 g

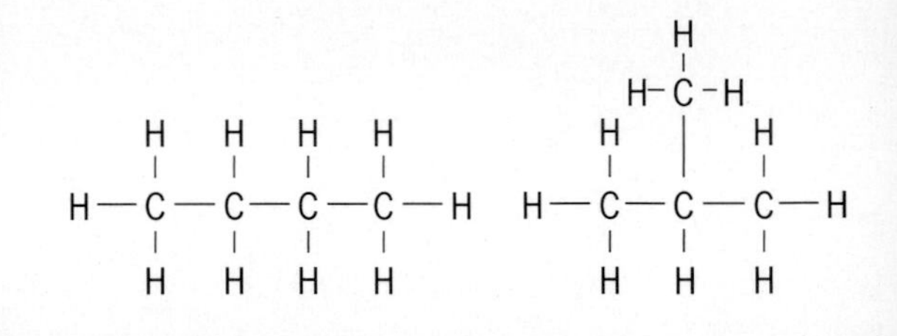

Add soap. A scum will be seen with hard water, a lather will be seen with soft water.

sodium hydroxide, hydrogen and chlorine

EXAM TECHNIQUE

FOLLOW OUR CHECKLIST TO HELP YOU BEFORE AND DURING THE EXAMS

Preparation

Use the time before the exams effectively. Write a list of all the topics you have to cover. Work through your notes systematically and ask for help with any topics that you're struggling to understand.

Practice

Attempt as many practice questions and past papers as possible. Familiarise yourself with the question types, the marks allocated and the time allowed. Compare your marks to those given in the mark schemes – see where you did well and where there is room for improvement.

Think positive

Even if time is running short, remind yourself of the progress you have made. Use what time is left by working through the key topics – either those that are most likely to come up in the exam or those that you find most difficult.

IN THE EXAM ITSELF...

- Follow all the instructions in the exam paper
- Attempt the correct number of questions
- Read each question carefully and more than once

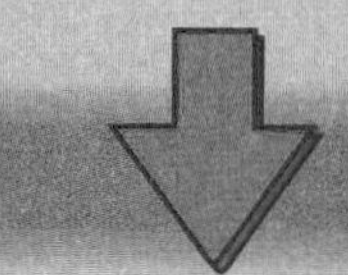

- Highlight the key words in the question and note the command word – State, Describe, Explain, Discuss, Find, Suggest, Calculate, List etc.
- Check the number of marks available for each question and answer accordingly

- Plan your response in brief note form
- Ensure that you answer the question asked and that your response stays relevant
- Allocate time carefully and make sure you complete the paper

- Return to any questions you have left out and read through your answers at the end
- Remember that accurate spelling and good use of English do count

We hope this book will help you on the way to GCSE success.

TITRATIONS 1

How to carry out a titration

Some specific apparatus is used in a titration. A **pipette** is used to measure accurately a known volume of one of the reactants (usually the alkali). This is then placed in a conical flask. The other reactant (usually the acid) is placed into a burette.

> **A burette allows you to add a liquid accurately until the exact point of neutralisation has occurred (called the end point).**

The end point is determined by the colour change of an **indicator**.

Titration apparatus

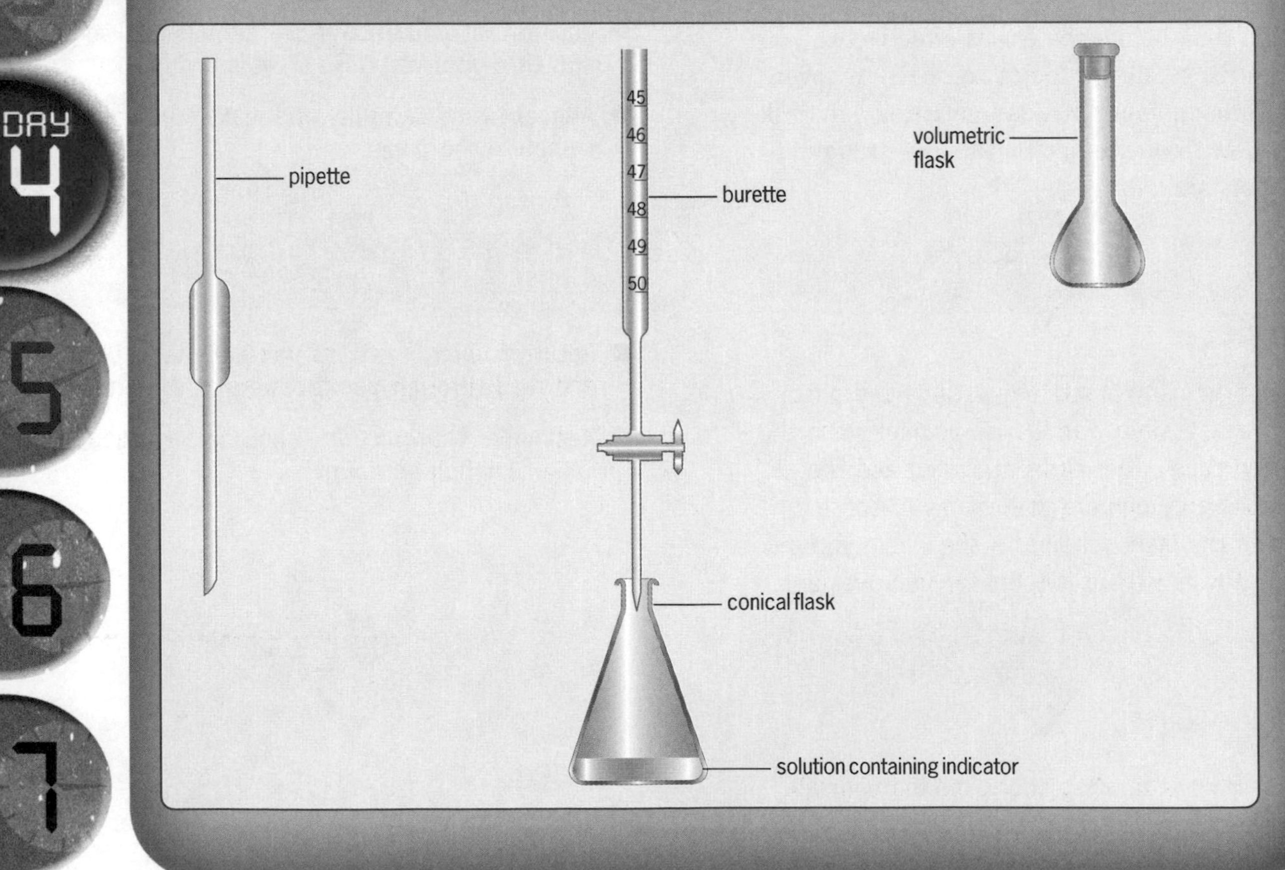

A titration is an accurate practical procedure that is used to identify either the volume or concentration of one of the reactants in an acid–alkali neutralisation.

What can a titration tell us?

During a titration an acid neutralises an alkali. We can use the results to find one of the following if you know the other three:

- volume of the acid
- concentration of the acid
- volume of the alkali
- concentration of the alkali

When doing titration calculations you need to be able to work out the number of moles (of acid and/or alkali) in solution.

You can use the following formula to do this.

$$\text{Number of moles} = \frac{\text{volume (cm}^3) \times \text{concentration (mol dm}^{-3})}{1000}$$

Example

How many moles of hydrochloric acid are there in 25 cm^3 of 0.12 mol dm^{-3} sodium hydroxide solution?

$$\text{Number of moles} = \frac{25 \times 0.12}{1000}$$

$$= 0.003\ (3 \times 10^{-3})$$

Progress check

1. Insert the missing words into the passage below.

   To measure out exactly 25 cm^3 of dilute alkali a __________ should be used. The liquid is then transferred to a clean __________ and a few drops of an __________ are added. The acid is then placed into a __________ and slowly added to the alkali until a __________ is observed.

2. Calculate:
   i. the number of moles in 12.5 cm^3 of 2 mol dm^{-3} hydrochloric acid
   ii. the concentration of 30 cm^3 of a solution containing 0.024 moles of sodium hydroxide

TITRATIONS 2

Rules for titration calculations

STEP 1 Calculate the number of moles of the substance of which you know the volume and concentration.

Remember Moles = $\frac{\text{volume} \times \text{concentration}}{1000}$

STEP 2 Use the equation for the reaction to work out the number of moles of the substance you are asked to find out about.

STEP 3 Rearrange the above formula to work out either the concentration or the volume of the substance you are asked to find out about.

Example

25 cm^3 of 0.13 mol dm^{-3} sodium hydroxide were neutralised by 17.90 cm^3 of hydrochloric acid. Calculate the concentration of the hydrochloric acid solution.

HCl (aq) + NaOH (aq) → NaCl (aq) + H2O (l)

STEP 1	You know the volume and concentration of the sodium hydroxide solution so you can work out the number of moles of sodium hydroxide used.	moles = volume × concentration ÷ 1000 $= \frac{25 \times 0.13}{1000} = 0.00325$
STEP 2	The ratio of sodium hydroxide to hydrochloric acid is one to one. 1HCl (aq) + 1NaOH (aq) → NaCl (aq) + H2O (l) This means that the number of moles of hydrochloric acid reacting will be exactly the same as the number of moles of sodium hydroxide reacting.	Number of moles of hydrochloric acid reacting is 0.003 25
STEP 3	The hydrochloric acid volume was 17.90 cm^3. You can now rearrange the formula to calculate the concentration of the hydrochloric acid solution. concentration = number of moles × 1000 ÷ volume	Concentration $= \frac{0.003\,25 \times 1000}{17.90}$ = 0.18 mol dm^{-3} (2 d.p.)

This chapter shows you how to solve titration problems. Like all of the other mathematical topics in chemistry, practice makes perfect!

Example

Calculate the volume of 0.56 mol dm^{-3} lithium hydroxide solution needed to neutralise 20 cm^3 of 0.36 mol dm^{-3} nitric acid.

HNO_3 (aq) + LiOH (aq)
→ $LiNO_3$ (aq) + H_2O(l)

STEP 1	Number of moles of nitric acid $= \frac{20 \times 0.36}{1000}$ = 0.0072
STEP 2	The reacting ratio is 1:1 so the number of moles of lithium hydroxide reacting is 0.0072 mol dm^{-3}.
STEP 3	Volume $= \frac{\text{number of moles} \times 1000}{\text{concentration}}$ $= \frac{0.0072 \times 1000}{0.56}$ = 12.86 cm^3 (2 d.p.)

Progress check

A student carried out a titration to work out the concentration of an unknown solution of sodium hydroxide. He titrated 25 cm^3 of the sodium hydroxide solution with 0.1 mol dm^{-3} hydrochloric acid. The amount of hydrochloric acid that was needed to neutralise the sodium hydroxide was 22.40 cm^3.

The equation for the reaction is:

HCl (aq) + NaOH (aq) → NaCl (aq) + H_2O (l)

1. Calculate the number of moles of hydrochloric acid reacting.

2. What is the ratio of moles of hydrochloric acid reacting to moles of sodium hydroxide reacting?

3. How many moles of sodium hydroxide were present in the 25 cm^3 of sodium hydroxide solution?

4. Calculate the concentration of the sodium hydroxide solution.

ATMOSPHERIC CHEMISTRY

1 2 3 4 DAY 5 6 7

Current composition of the atmosphere

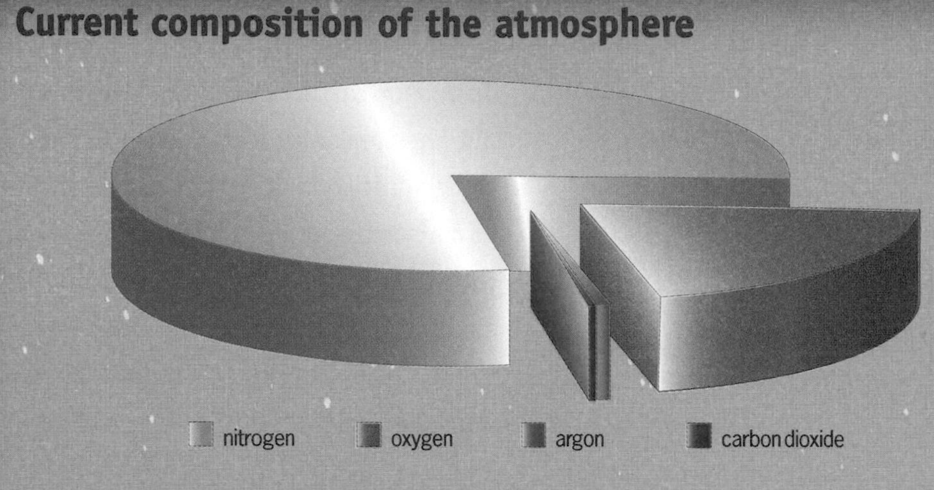

For 200 million years the proportion of different gases in the atmosphere has been much the same as they are today:

- about 80% nitrogen
- about 20% oxygen
- small proportion of various other gases, including carbon dioxide, water vapour and noble gases (mainly argon)

Evolution of the current atmosphere

During the first billion years of the Earth's existence there was intense volcanic activity. This activity released the gases which then formed the early atmosphere mainly **ammonia** (**NH_3**) and **methane** (**CH_4**) and **water** vapour that eventually cooled and condensed to form the oceans. During this period the Earth's atmosphere was probably mainly **carbon dioxide** and there would have been little or no oxygen gas. This is like the atmospheres of Mars and Venus today.

When plants evolved and successfully colonised most of the Earth's surface:

- The atmosphere became more and more 'polluted' with **oxygen**. This means that there were fewer habitats suitable for micro-organisms, which could not tolerate oxygen.
- Most of the carbon from the carbon dioxide in the air gradually became locked up in sedimentary rocks such as carbonates and fossil fuels.
- The methane and ammonia in the atmosphere reacted with the oxygen.

The composition of the atmosphere is very different today from how it was when the planet was first formed. The chemistry that occurs in the environment has turned the original atmosphere into the one that we have today.

- Nitrogen gas was released in to the air, partly from the reaction between oxygen and ammonia, but mainly from living organisms, including **denitrifying bacteria**.
- The oxygen in the atmosphere resulted in the development of the **ozone layer**. This filters out harmful ultraviolet radiation from the sun, allowing the evolution of new living organisms.

Carbon dioxide in the atmosphere

Carbonate rocks are sometimes moved deep into the Earth by geological activity. They may then release carbon dioxide back into the atmosphere via volcanoes.

The release of carbon dioxide by burning carbon locked up over hundreds of millions of years in fossil fuels increases the level of carbon dioxide in the atmosphere. Though the reaction between carbon dioxide and sea-water increases, producing insoluble (mainly calcium) carbonates which are deposited as sediment and soluble hydrogencarbonates (mainly calcium and magnesium), this does not wholly absorb the additional carbon dioxide released into the atmosphere.

Progress check

1. What is the approximate percentage of nitrogen in the atmosphere today?
   a) less than 1%
   b) 20%
   c) 80%
   d) 99%
2. Name one gas that was present in the early atmosphere that is not present in today's atmosphere.
3. Explain how oxygen gas became present in the atmosphere.
4. Give two ways in which carbon dioxide is added to the atmosphere.
5. Give two ways in which carbon dioxide is removed from the atmosphere.

THE ROCK CYCLE

This diagram gives information about the different stages of the rock cycle.

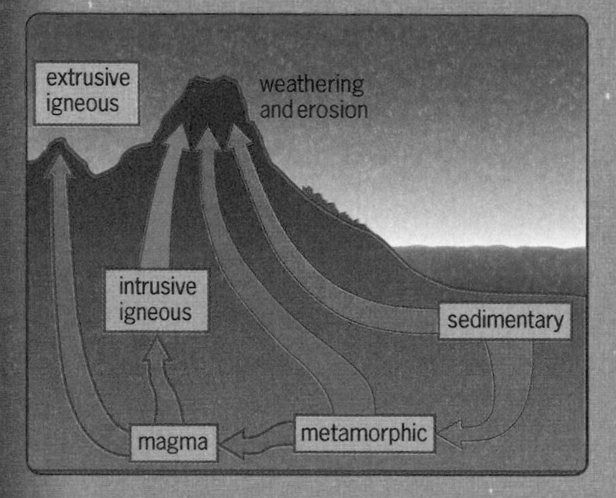

Igneous rocks

Igneous rocks are formed by the **cooling** and **solidifying** (crystallisation) of molten magma. Magma that reaches the surface of the Earth cools and solidifies more rapidly than magma that cools inside the Earth.

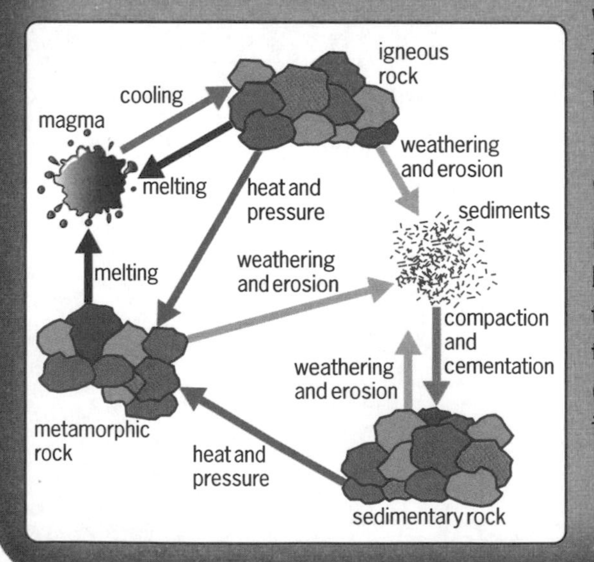

Intrusive igneous rocks (e.g. granite) are formed when magma crystallises inside the Earth's crust. **As crystallisation occurs slowly the crystals present are large.**

Extrusive igneous rocks (e.g. basalt) are formed when magma crystallises on the Earth's surface. **As crystallisation occurs quickly the crystals present are small.**

Weathering and erosion

Weathering is the breakdown of stationary rock. Methods of weathering include **physical**, e.g. freeze–thaw, **chemical** (acid rain etc.) and **biological** (plant root growth etc.).

Erosion occurs when broken-off bits of rock (from weathering) are further broken down by movement, e.g. by gravity as rock fragments fall down a hill or by water in streams, rivers etc.

Weathering and erosion over long periods of time turn igneous rock into small particles of rock called **sediments**.

Sedimentary rock

Eventually these sediments end up at the bottom of the sea. The **high pressure** of the sea over long periods of time compacts the sediments into sedimentary rock (e.g. limestone, mudstone). This process is called **cementation**.

The rock cycle shows how the three main types of rock are constantly interchanging.

Sediments contain evidence of how they were deposited (e.g. layers formed by discontinuous deposition, ripple marks formed by currents or waves). At the surface of the Earth younger sedimentary rocks usually lie on top of older rocks.

Sedimentary rocks are often found tilted, folded, fractured (faulted) and sometimes even turned upside down. This shows that **the Earth's crust is unstable** and has been subjected to very large forces.

Large-scale **movements of the Earth's crust can cause mountain ranges** to form very slowly over millions of years. These replace older mountain ranges worn down by weathering and erosion.

Metamorphic rock

Metamorphic rock (e.g. marble, slate, schist) is formed by the effects of **heat** and **pressure** recrystallising other (usually sedimentary) rocks.

Metamorphic rocks are associated with the Earth's movements (**tectonic activity**) which created present-day and ancient mountain belts. They are evidence of the high temperatures and pressure created by these mountain-building processes.

Progress check

1. Name the three different types of rock.
2. What is the main difference in appearance between intrusive and extrusive igneous rock?
3. What causes this difference to occur?
4. Give an example of each of biological, chemical and physical weathering.
5. What two conditions are needed for metamorphic rock to be formed?

THE PERIODIC TABLE AND THE TRANSITION ELEMENTS

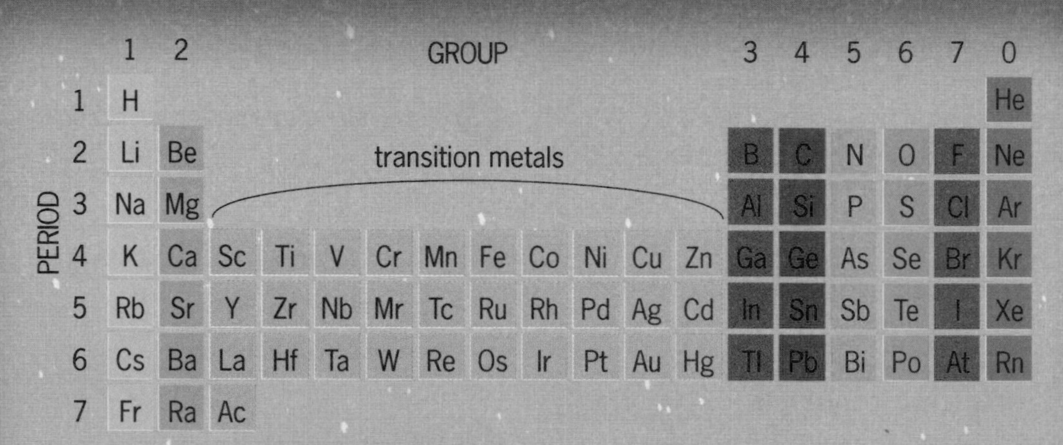

PERIOD	1	2											3	4	5	6	7	0
1	H																	He
2	Li	Be					transition metals						B	C	N	O	F	Ne
3	Na	Mg											Al	Si	P	S	Cl	Ar
4	K	Ca	Sc	Ti	V	Cr	Mn	Fe	Co	Ni	Cu	Zn	Ga	Ge	As	Se	Br	Kr
5	Rb	Sr	Y	Zr	Nb	Mr	Tc	Ru	Rh	Pd	Ag	Cd	In	Sn	Sb	Te	I	Xe
6	Cs	Ba	La	Hf	Ta	W	Re	Os	Ir	Pt	Au	Hg	Tl	Pb	Bi	Po	At	Rn
7	Fr	Ra	Ac															

GROUP

The organisation of the periodic table

The chemical elements can be arranged in order of their relative atomic masses. This list can be arranged in rows so that **elements with similar properties are in the same columns** known as **groups**. The resulting table is known as the **periodic table**. However, if the elements are ordered according to relative atomic mass a few elements will not end up in the correct group e.g. Te and I.

The modern periodic table is arranged in order of the atomic (proton) number. All elements are then in the appropriate groups.

The periodic table can be seen as an arrangement of the elements in terms of their electronic structure. From left to right, across each horizontal row (**period**) of the periodic table, a particular energy level is gradually filled up with electrons.

In the next period, the next shell is filled with electrons.

The similarities and differences between the properties of elements in the same group of the periodic table can be explained by the electronic structure of their atoms. More than three quarters of the elements are metals. They are mainly found:

- in the two left-hand columns (group 1 and group 2)
- in the central block (transition elements)

Fewer than one quarter of the elements are non-metals. Non-metal elements are found in the groups at the right-hand side of the periodic table.

Elements in the same group have similar properties because they have the same number of electrons in the highest occupied (outer) energy level (shell).

It is important for you to understand the trends and patterns in the periodic table if you are to understand the chemical reactions of the elements.

Transition metals

In the centre of the periodic table is a block of metallic elements. These elements, which include iron and copper, are known as **transition metals**.

Like all other metals, transition metals are good conductors of heat and electricity and can easily be hammered or bent into shape.

Compared to alkali metals, transition metals:

- have **high melting points** (except for mercury which is a liquid at room temperature)
- are **hard**, **tough** and **strong**
- are **much less reactive** and so do not react (corrode) as quickly with oxygen and/or water

These properties make transition metals very useful as **structural materials** (e.g. iron, usually in the form of steel) and for making things that must allow heat or electricity to pass through them easily (e.g. copper for electrical cables).

Most transition metals form **coloured compounds**. These can be seen:

- in pottery glazes of various colours
- in weathered copper (green)

Many transition metals are used as **catalysts**, e.g. iron in the Haber process and platinum in the manufacture of nitric acid.

Progress check

1. How are elements arranged in the periodic table?
2. What name is given to rows of elements?
3. Why are elements arranged in groups?
4. Give two ways in which the transition metals differ from group 1 metals.
5. Give a use for a named transition metal.

METALS: GROUP 1

Group 1 elements –the alkali metals

The alkali metals:

- are metals with a low density (the first three are less dense than water and therefore float on water)
- react with non-metals to form ionic compounds in which the metal ion will have a charge of +1
- form compounds that are white solids and which dissolve in water to form colourless solutions
- react with water releasing hydrogen
- form hydroxides (OH) that dissolve in water to give alkaline solutions

Reactivity of group 1 metals

In group 1, the further down the group an element is:

- **the more reactive the element**
- **the lower the melting point and boiling point**

Reaction with water

When a piece of lithium, sodium or potassium is placed in cold water the metal floats, it may melt and it moves around the surface of the water. Fizzing and bubbling (**effervescence**) will occur. With potassium, a flame is seen. The metal reacts with the water to form a metal hydroxide solution and hydrogen gas.

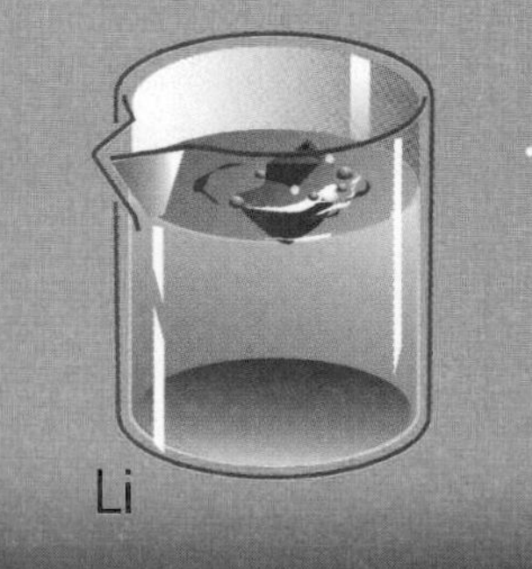

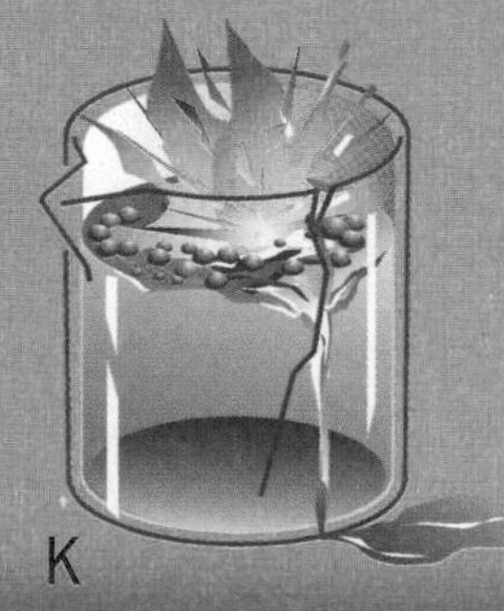

The group 1 elements (alkali metals) are interesting metals because of their high reactivity. It is also important to understand the trend in their reactivity.

alkali metal + water
→ metal hydroxide + hydrogen

For example, using sodium as an example:

$2Na\ (s) + 2H_2O\ (l) \rightarrow 2NaOH\ (aq) + H_2\ (g)$

The more reactive the metal, the more vigorous is the reaction with water.

Reactivity increases as you go down group 1 because:

The atoms get larger.

↓

Therefore the outer electron becomes further away from the nucleus.

↓

Therefore there is less attraction between the nucleus and the electron.

↓

Therefore it becomes easier for the atom to lose the outer electron and hence it is more reactive.

A simple laboratory test for hydrogen is that when a test-tube of hydrogen is held to a flame the hydrogen burns in the air, with a squeaky pop.

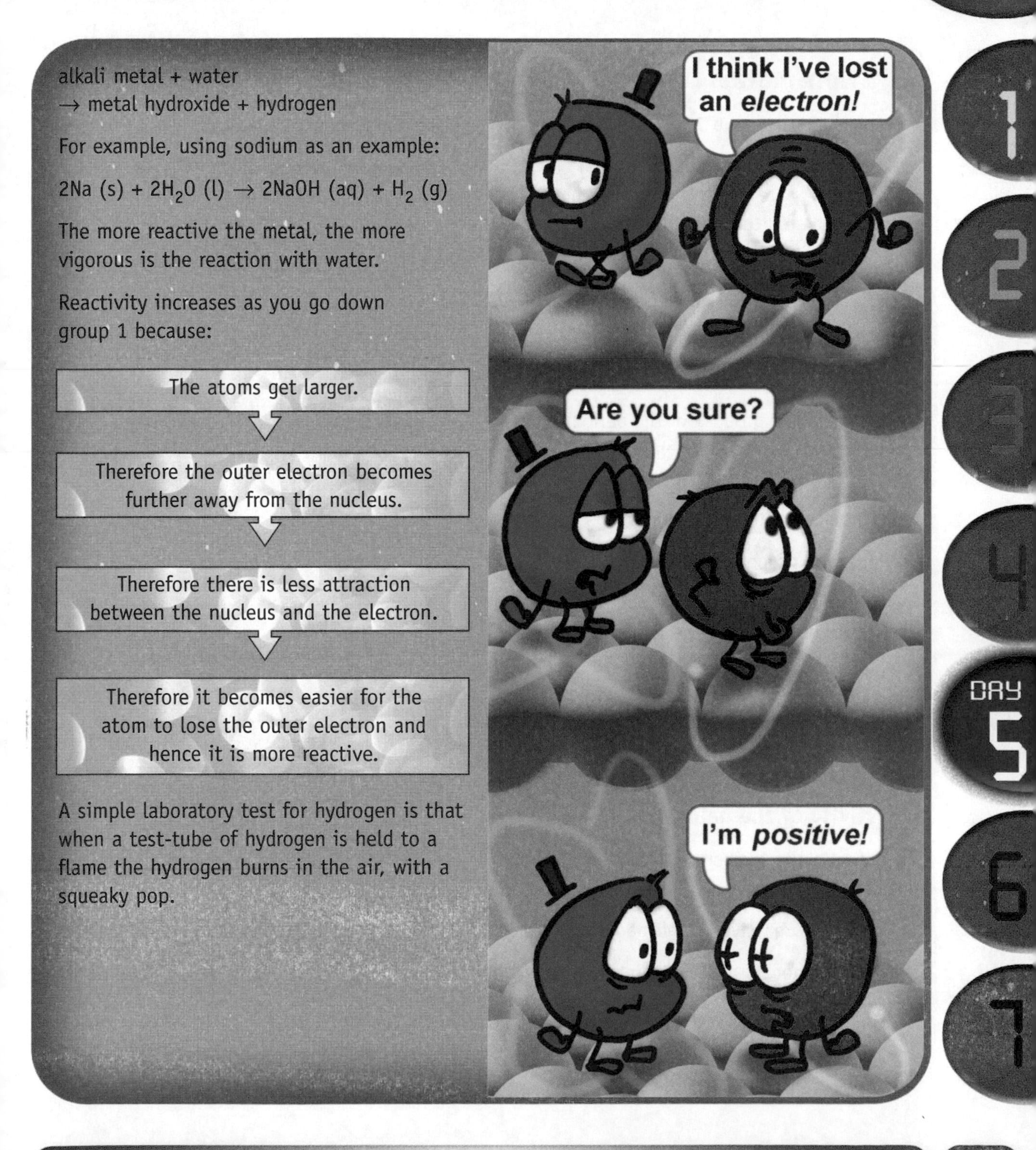

NON-METALS: GROUPS 7 AND 0

Non-metals

The elements in group 7 and group 0 have the typical properties of non-metals.

- They have low melting points and boiling points. At room temperature all the group 0 elements are gases, the first two group 7 elements are gases, the third (bromine) is a liquid and the fourth (iodine) is a solid.
- They are brittle and crumbly when solid.
- They are poor conductors of heat and electricity even when solid or liquid.

Group 7 – the halogens

The elements in group 7 of the periodic table (known as the halogens):

- have coloured vapours (chlorine is green, bromine is red, iodine is purple)
- consist of molecules that are made up of pairs of atoms
- form ionic salts with metals, in which the chloride, bromide and iodide ions (halide ions) carry a charge of –1
- form molecular compounds with other non-metallic elements

Reactivity of the halogens

In group 7, the further down the group an element is:

- the less reactive the element
- the higher its melting point and boiling point

A more reactive halogen can displace a less reactive halogen from an aqueous solution of its salt.

e.g. chlorine + potassium iodide
→ potassium chloride + iodine

$Cl_2\ (g) + 2KI\ (aq) \rightarrow 2KCl\ (aq) + I_2$

Reactivity increases as you go up group 7 because:

The atoms get smaller.

↓

Therefore the outer shell is closer to the nucleus.

↓

Therefore there is stronger attraction between the nucleus and the electron in another atom.

↓

Therefore the atom is more able to attract that electron and so it is more reactive.

The non-metals have the opposite trend in reactivity to the metals, i.e. they get more reactive as you go up the group.

Group 0 – the noble gases

Group 0 elements:

- are all chemically very unreactive gases
- exist as individual atoms rather than diatomic gases (like most other gaseous elements)
- are used as inert gases in filament lamps and in electrical discharge tubes (e.g. fluorescent lighting)

The first element in the group, helium, is much less dense than air and is used in balloons.

Reactivity of the noble gases

Group 0 elements are unreactive and monatomic because their outer shell of electrons is full, so atoms have no tendency to gain, lose or share electrons.

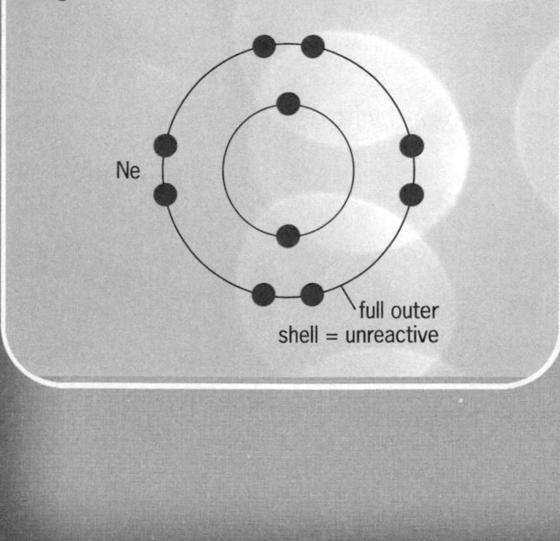

Progress check

1. Give two typical properties of non-metals.

2. Write true or false after each of the following statements.
   i. Chlorine gas has a yellow/green colour.
   ii. The halogens exist as single atoms.
   iii. Chlorine is less reactive than bromine.
   iv. Noble gases exist as single atoms.
   v. Noble gases are more reactive than the halogens.

3. Explain why neon is such an unreactive gas.

4. Complete the following equation.
   Sodium iodide + bromine →

5. Explain why bromine is more reactive than iodine.

THE ELECTROLYSIS OF BRINE

Electrolysis of sodium chloride solution (brine)

Sodium chloride (common salt) is a compound of an alkali metal and a halogen. It is found in large quantities in the sea and in underground deposits. The electrolysis of sodium chloride solution is an important industrial process.

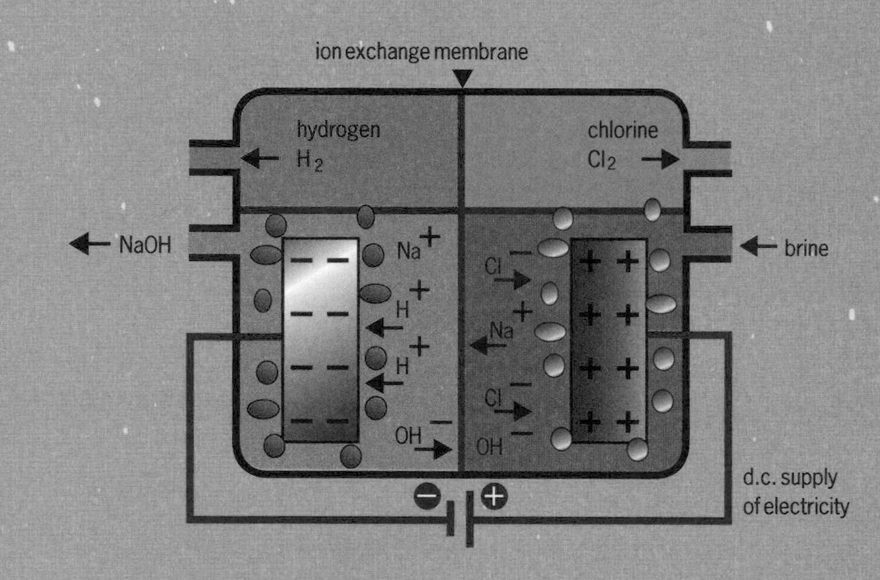

The overall equation for this reaction is:

$$\text{sodium chloride} + \text{water} \xrightarrow{\text{electricity}} \text{hydrogen} + \text{chlorine} + \text{sodium hydroxide}$$

$$2NaCl + 2H_2O \rightarrow H_2 + Cl_2 + 2NaOH$$

The reactions occurring at each electrode are:

Cathode (negative electrode)	Anode (positive electrode)
$2H^+ + 2e^- \rightarrow H_2$	$2Cl^- \rightarrow Cl_2 + 2e^-$

Brine is concentrated sodium chloride solution. Electrolysis of this produces three useful products that can be used for other chemical processes.

The electrodes are made of an unreactive metal – usually titanium. Titanium is used because it is a good conductor of electricity and it will not react with the chlorine gas or sodium hydroxide solution.

The products

Chlorine gas is formed at the positive electrode (anode) and **hydrogen** gas at the negative electrode (cathode). The resulting solution is **sodium hydroxide**. These three products are used to make other useful materials.

Substance	Use
Chlorine	To kills bacteria in drinking water and swimming pools Manufacture of hydrochloric acid, disinfectants, bleach and PVC plastic
Hydrogen	Manufacture of ammonia and margarine
Sodium hydroxide	Manufacture of soap, paper and ceramics

A simple laboratory **test for chlorine** is that it **bleaches damp litmus paper**.

Progress check

1. Name the three products formed during the electrolysis of brine.
2. Complete the diagram to label the anode (positive electrode) and cathode (negative electrode).

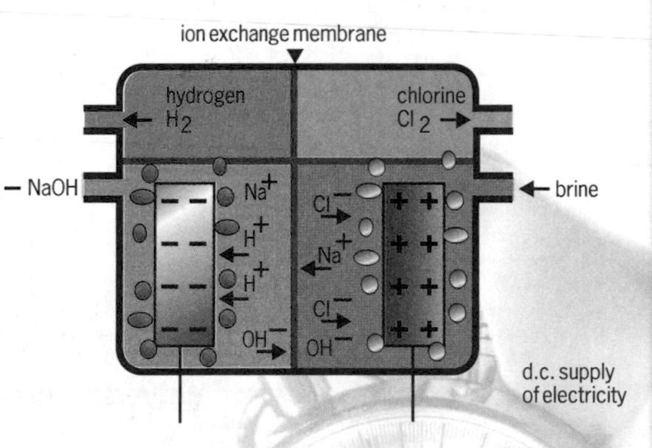

3. What is the chemical test and result for chlorine gas?
4. Give one use for each of the products in this process.
5. Explain why titanium is used for the electrodes in this process.

Acidity and indicators

When a substance dissolves in water it forms an **aqueous solution** which may be **acidic**, **alkaline** or **neutral**. Water itself is neutral.

Indicators can be used to show whether a solution is acidic, alkaline or neutral by the way their colours change.

The pH scale is used to show how acidic or alkaline a solution is.

pH	1	2	3	4	5	6	7	8	9	10	11	12	13	14
Colour	Red		Orange		Yellow			Green			Blue		Violet	
Strength	strong ACIDS ←				weak		neutral	weak			→ ALKALIS strong			

The colours show the colour of **universal indicator** (UI) in a solution of that pH.

An indicator can be used to show when acidic and alkaline solutions have completely reacted to form a neutral salt solution.

Salts

Compounds of alkali metals called **salts** can be made by reacting solutions of their hydroxides (which are alkaline) with acids. These are **neutralisation reactions**.

acid + metal hydroxide solution → neutral salt solution + water

For example:

hydrochloric acid + sodium hydroxide → sodium chloride + water

The particular salt produced in any reaction depends on:

- the acid used
- the metal used in the alkali

Many chemical reactions involve the reactions of acids and alkalis. When an acid is reacted with an alkali, a neutralisation reaction occurs.

Neutralising **hydrochloric acid** produces **chlorides**.

Example

lithium hydroxide + hydrochloric acid → lithium chloride + water

$LiOH + HCl \rightarrow LiCl + H_2O$

Neutralising **nitric acid** produces **nitrates**.

Example

sodium hydroxide + nitric acid → sodium nitrate + water

$NaOH + HNO_3 \rightarrow NaNO_3 + H_2O$

Neutralising **sulphuric acid** produces **sulphates**.

Example

potassium hydroxide + sulphuric acid → potassium sulphate + water

$2KOH + H_2SO_4 \rightarrow K_2SO_4 + 2H_2O$

Ammonia (NH_3) also dissolves in water to produce an alkaline solution. This can be neutralised with acids to produce **ammonium salts**.

Example

ammonia solution + sulphuric acid → ammonium sulphate

$2NH_3 + H_2SO_4 \rightarrow (NH_4)_2SO_4$

Progress check

1. What colour will universal indicator turn if it is placed in a neutral solution?

2. What pH does a neutral solution have?

3. Complete these equations to show the reactants/products of some neutralisation reactions.
   i. hydrochloric acid + lithium hydroxide →
   ii. sulphuric acid + calcium hydroxide →
   iii. nitric acid + sodium hydroxide →

4. Suggest the acid and alkali that could be used to make these salts.
   i. potassium chloride
   ii. lithium sulphate
   iii. calcium nitrate

ACIDS 2

DAY 6

Ions and neutralisation

In any neutralisation reaction the following ionic reaction occurs.

$H^+ (aq) + OH^- (aq) \rightarrow H_2O (l)$

Hydrogen ions, H^+ (aq), make solutions acidic.

Hydroxide ions, OH^- (aq), make solutions alkaline.

Alkalis and bases

Alkalis are soluble bases. Both bases and alkalis react with acids but bases, e.g. copper carbonate, are not soluble in water.

Reactions of acids

There are four different types of reactions of acids.

1 acid + alkali → salt + water

For example: sulphuric acid + sodium hydroxide → sodium sulphate + water

$H_2SO_4 + 2NaOH \rightarrow Na_2SO_4 + 2H_2O$

2 acid + metal → salt + hydrogen

For example: hydrochloric acid + magnesium → magnesium chloride + hydrogen

$2HCl + Mg \rightarrow MgCl_2 + H_2$

3 acid + metal carbonate → salt + carbon dioxide + water

For example: nitric acid + calcium carbonate → calcium nitrate + carbon dioxide + water

$2HNO_3 + CaCO_3 \rightarrow Ca(NO_3)_2 + CO_2 + H_2O$

4 acid + metal oxide → salt + water

For example: sulphuric acid + copper(II) oxide → copper(II) sulphate + water

$H_2SO_4 + CuO \rightarrow CuSO_4 + H_2O$

Soluble and insoluble salts

Soluble salts	Insoluble salts
All sodium, potassium and ammonium salts	
All nitrates	
Chlorides	Except silver and lead(II) chloride
Sulphates	Except calcium, barium and lead(II) sulphate
Sodium, potassium and ammonium carbonates	All other carbonates are insoluble

Preparing an insoluble salt

Not all salts are soluble. Insoluble salts are made by a **precipitation reaction** where an insoluble solid is made by mixing together two different solutions.

This section describes the ions that make substances acid or alkaline and what happens during neutralisation reactions.

Preparing a soluble salt

A salt can be obtained from reacting an acid with an alkali by means of a titration procedure (see page 48). To obtain the salt the titration would have to be repeated without an indicator and the neutral solution would need to be heated to evaporate off all the water to leave behind the salt crystals.

It is easier to prepare a salt by reacting an acid with a metal or base, e.g. preparing copper(II) sulphate from copper(II) oxide and sulphuric acid.

1 Add some copper(II) oxide to some sulphuric acid and gently warm the mixture (to speed up the rate of the reaction).

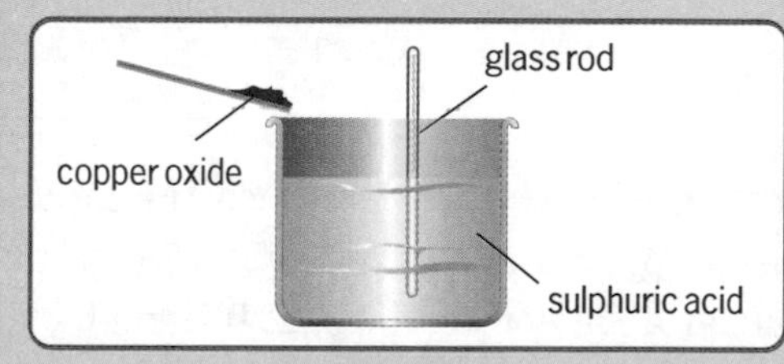

2 Stir the mixture and keep on adding copper(II) oxide until no more dissolves. Filter the mixture.

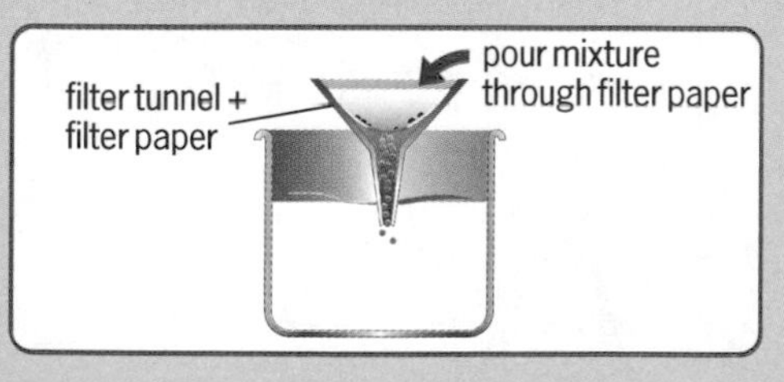

3 Heat the filtrate (or leave it in a warm place) to evaporate off the water, leaving the crystals of copper(II) sulphate behind.

Progress check

1. Which ion is present in all acids?
2. Write an ionic equation for the reaction of hydrochloric acid with sodium hydroxide.
3. Complete the blanks to show the missing reactants/products in these reactions.
   i. sulphuric acid + magnesium → ________ + ________
   ii. hydrochloric acid + ________ → calcium chloride + water
   iii. ________ + ________ → copper(II) sulphate + carbon dioxide + water
4. Describe how you would prepare crystals of zinc sulphate, starting from zinc metal and sulphuric acid.

RATES OF REACTION 1

Factors affecting rates

The speed (rate) of a chemical reaction increases:

- if the temperature increases
- if the concentration of dissolved reactants or the pressure of gases increases
- if solid reactants are in smaller pieces (greater surface area)
- if a catalyst is used

Catalysts

A catalyst increases the rate of a chemical reaction without being used up during the reaction.

Catalysts work by lowering the activation energy of the reactant(s) and this allows the reaction to happen more easily.

A catalyst is used over and over again to speed up the conversion of reactants to products. Different reactions need different catalysts.

Following the rate of a reaction

The rate of a chemical reaction can be followed by measuring the rate at which the products are formed or the rate at which the reactants are used up. This allows a comparison to be made of the changing rate of a chemical reaction under different conditions. Different methods of following the rate of a reaction include collecting gas produced at different time intervals, measuring mass loss (e.g. when a gas is lost during a reaction) and measuring colour changes.

The apparatus in the diagram is used to collect the volume of gas produced when magnesium metal reacts with hydrochloric acid.

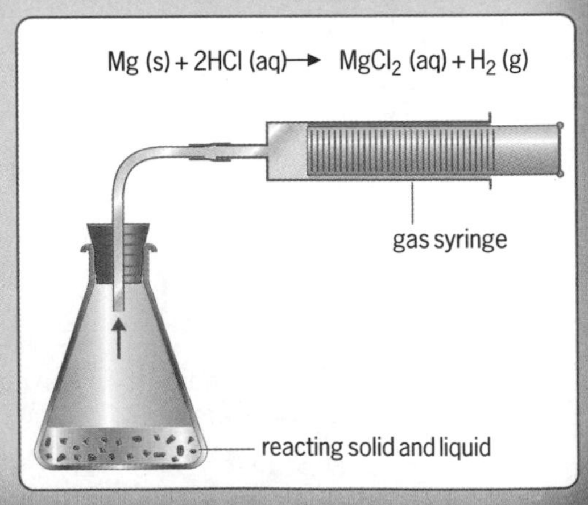

In an experiment the gas was collected every minute for 15 minutes. The experiment was carried out four times, each time with a different concentration of acid. The results are shown on the following graph.

Increasing the rates of chemical reactions is important in industry because it helps to reduce costs.

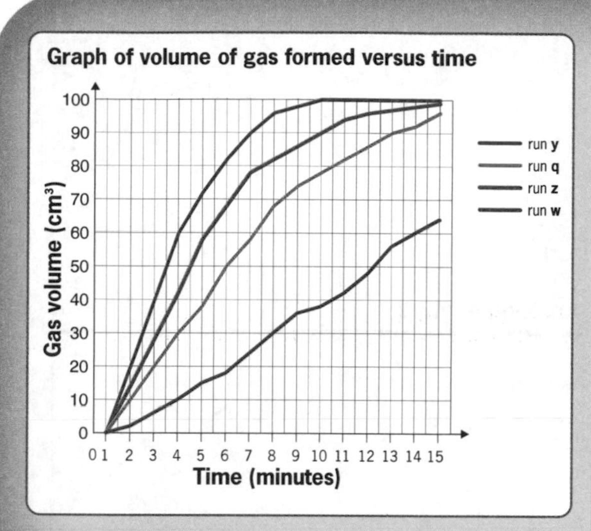

Interpretation

The green line (run w) represents the reaction where the concentration of acid was the highest. The gradient of the initial part of the curve is the steepest. This means that the rate of the reaction is fastest there. The red line (run y) represent the reaction where the concentration of acid was the lowest.

The final volume of gas was always 100 cm^3 because the reaction will always go to completion but just at different rates. Run w finished after 10 minutes.

The rate of all of the above reactions will slow down as the reaction proceeds. This is because the concentration of the acid will decrease, so there will be fewer collisions and thus the rate decreases.

Progress check

1. Give two ways of increasing the rate of a reaction.
2. Explain how a catalyst increases the rate of a reaction.
3. Suggest why the rate of reaction of hydrochloric acid with marble chips decreases as the reaction proceeds.

10 MINS RATES OF REACTION 2

Collision theory

Chemical reactions can only occur when reacting particles collide with each other and with sufficient energy. The minimum amount of energy particles must have to react is called the **activation energy**. If they collide with the activation energy then a successful collision occurs.

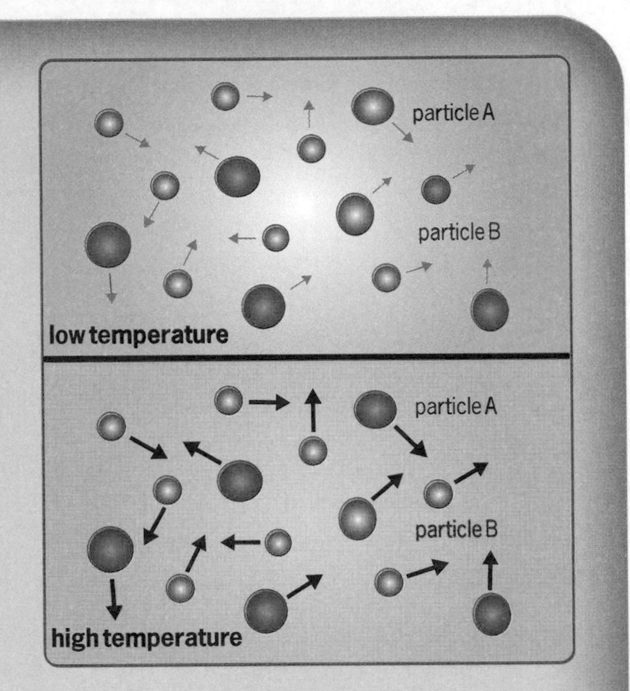

Collision theory and temperature

Increasing the temperature increases the speed of the reacting particles so that they collide more frequently and with more energy (i.e. more particles will possess the activation energy). This increases the rate of reaction.

Collision theory and concentration

Increasing the concentration of reactants in solutions and increasing the pressure of reacting gases also increases the frequency of collisions and so increases the rate of reaction.

The lower diagram shows a higher concentration of reactants and hence there are more collisions.

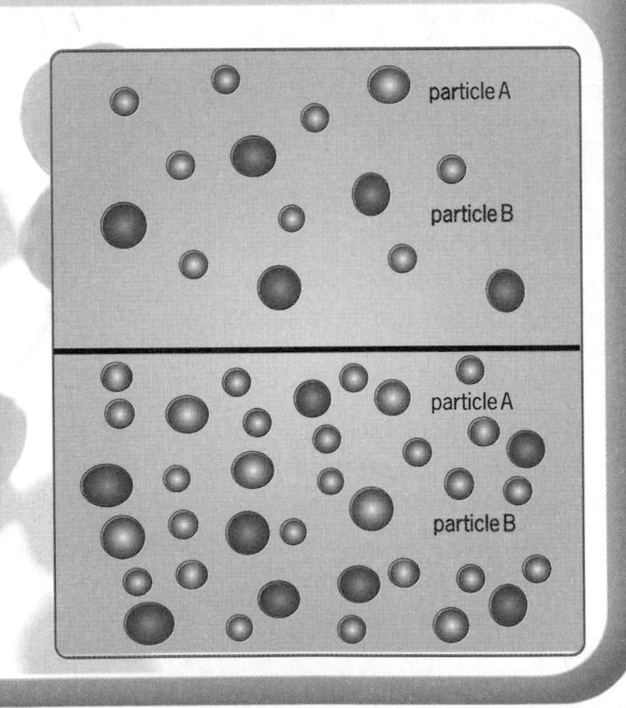

This topic looks at how the collision theory can be used to explain rates of reaction.

Collision theory and surface area

If one of the reactants in a reaction is a solid then the rate of reaction can be increased by increasing the surface area of the solid reactant. This provides more spaces where collisions can occur.

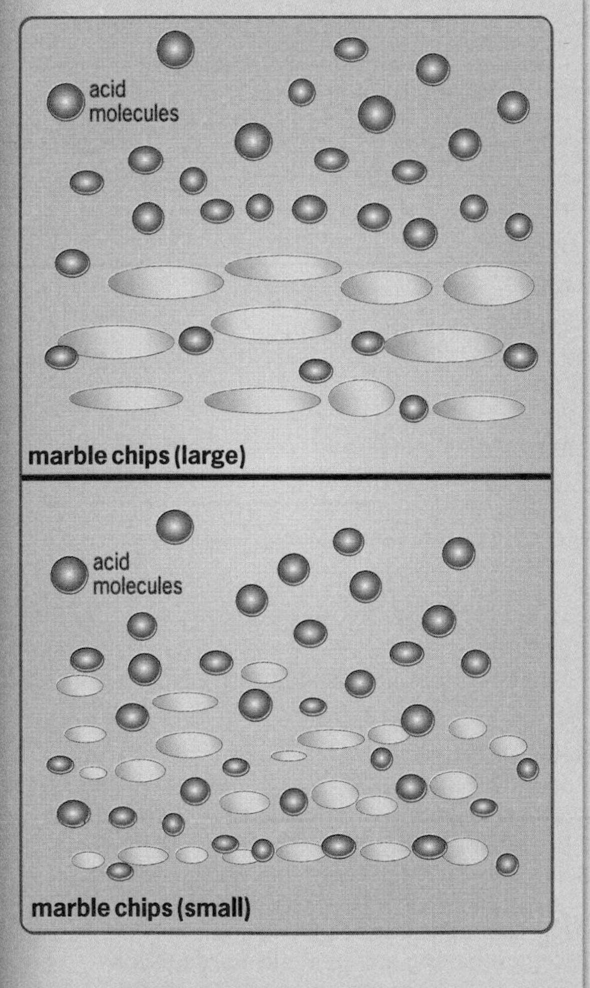

Progress check

1. What must particles possess if a successful collision is to occur?
2. Give two reasons why increasing the temperature increases the rate of a chemical reaction.
3. Explain why doubling the concentration of a reactant should theoretically double the rate of a reaction.
4. Why does using smaller marble chips in the reaction with hydrochloric acid result in a greater rate of reaction than if larger chips are used?

REACTIONS INVOLVING ENZYMES

Fermentation

Yeast cells convert sugar into carbon dioxide and ethanol.

sugar → alcohol + carbon dioxide

A simple laboratory test for carbon dioxide is that it turns limewater milky.

This process is called **fermentation** and is used:

- to produce the alcohol in beer and wine
- to produce the bubbles of carbon dioxide that make bread dough rise

Enzymes and temperature

The chemical reactions brought about by living cells are quite fast in conditions that are warm rather than hot. This is because the cells use catalysts called **enzymes**. **Enzymes are protein molecules that act as biological catalysts**. Different enzymes work best at different pH values.

Enzymes are usually **denatured** (damaged) by temperatures above about 45°C. The diagram on the right shows how rate of enzyme reaction is affected by temperature.

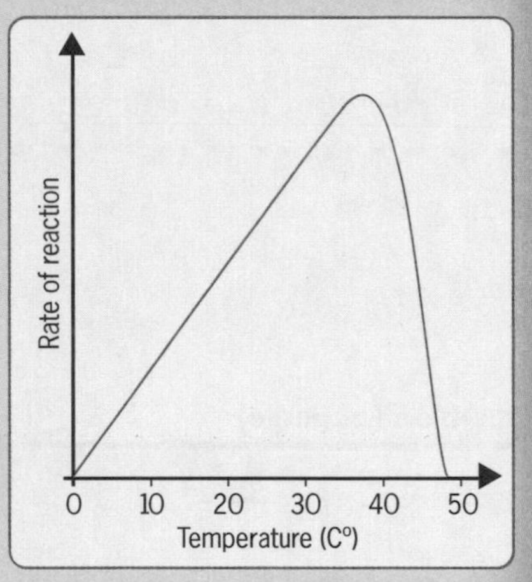

Uses of enzymes

In the home 	Bread dough raising Biological detergents may contain protein-digesting protease enzymes and fat-digesting lipase enzymes
In industry, enzymes are used to bring about reactions at normal temperatures and pressures that would otherwise require more expensive and more energy-demanding equipment	**Proteases** break down proteins and are used to 'pre-digest' the protein in some baby foods **Carbohydrases** are used to convert starch syrup into sugar and syrup **Invertase** is used to make the sugar for soft chocolates **Pectinase** breaks down insoluble pectin polysaccharides and so is used to clarify fruit juices **Amylases** break down carbohydrates and **lipases** break down fats Enzymes are used in **genetic engineering** and **penicillin production** The dairy industry uses enzymes made by micro-organisms (bacteria) to produce **yoghurt and cheese** from milk **Bacterial enzymes** convert the sugar in milk (lactose) to lactic acid Enzymes in **biological detergents** help break down staining food materials

Living cells use chemical reactions to produce new materials. They use enzymes (biological catalysts) to increase the rates of these reactions. Enzymes are also used widely in industry.

Successful industrial processes depending on enzymes usually:

- **stabilise the organism** to keep it functioning for a long period
- **immobilise the enzyme** by trapping it in an inert solid support or carrier such as alginate beads
- allow a continuous process: this means a continuous input of raw materials and output of product, so they can run 24 hours a day for many weeks or months (rather than a batch process, which means loading the reactor vessel with reactants, extracting the product, cleaning out, then re-loading with reactants and so on).

Progress check

1. Complete the following equation for fermentation.

   sugar → ethanol + __________

2. Explain why reactions involving enzymes should be kept at temperatures below 45°C.

3. Give one use of enzymes in the home.

4. Give one use of a named enzyme in industry.

Exothermic and endothermic energy changes

When fuels burn, energy is released as heat. Whenever chemical reactions occur, energy is usually transferred to or from the surroundings.

An **exothermic** reaction is one which transfers energy, often as heat, to the surroundings. An **endothermic** reaction is one which takes in energy, often as heat, from the surroundings.

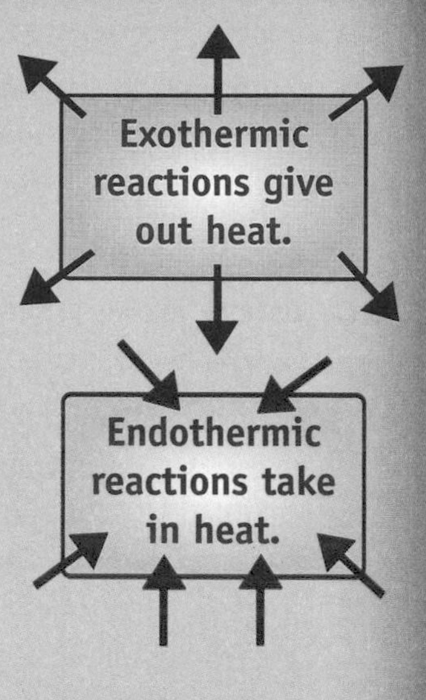

If a reversible reaction is exothermic in one direction it is endothermic in the opposite direction. The same amount of energy is transferred in each case.

Example

blue hydrated copper(II) sulphate + heat ⇌ white anhydrous copper(II) sulphate + water

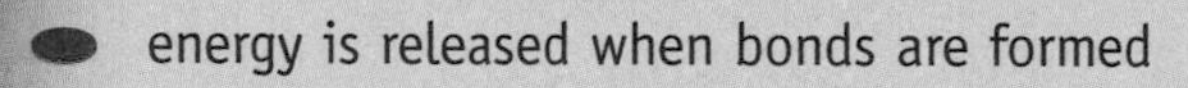

The reverse reaction can be used as a test for water, i.e. anhydrous copper sulphate + water. The white anhydrous copper sulphate turns blue if water is present.

During a chemical reaction:

- energy must be supplied to break bonds
- energy is released when bonds are formed

In an **exothermic reaction**, the **energy released** from forming new bonds is greater than the energy needed to break existing bonds.

In an **endothermic reaction**, the **energy needed** to break existing bonds is greater than the energy released from forming new bonds.

Most chemical reactions involve a change in energy. This energy is in the form of heat.

10 MINS

Energy level diagrams

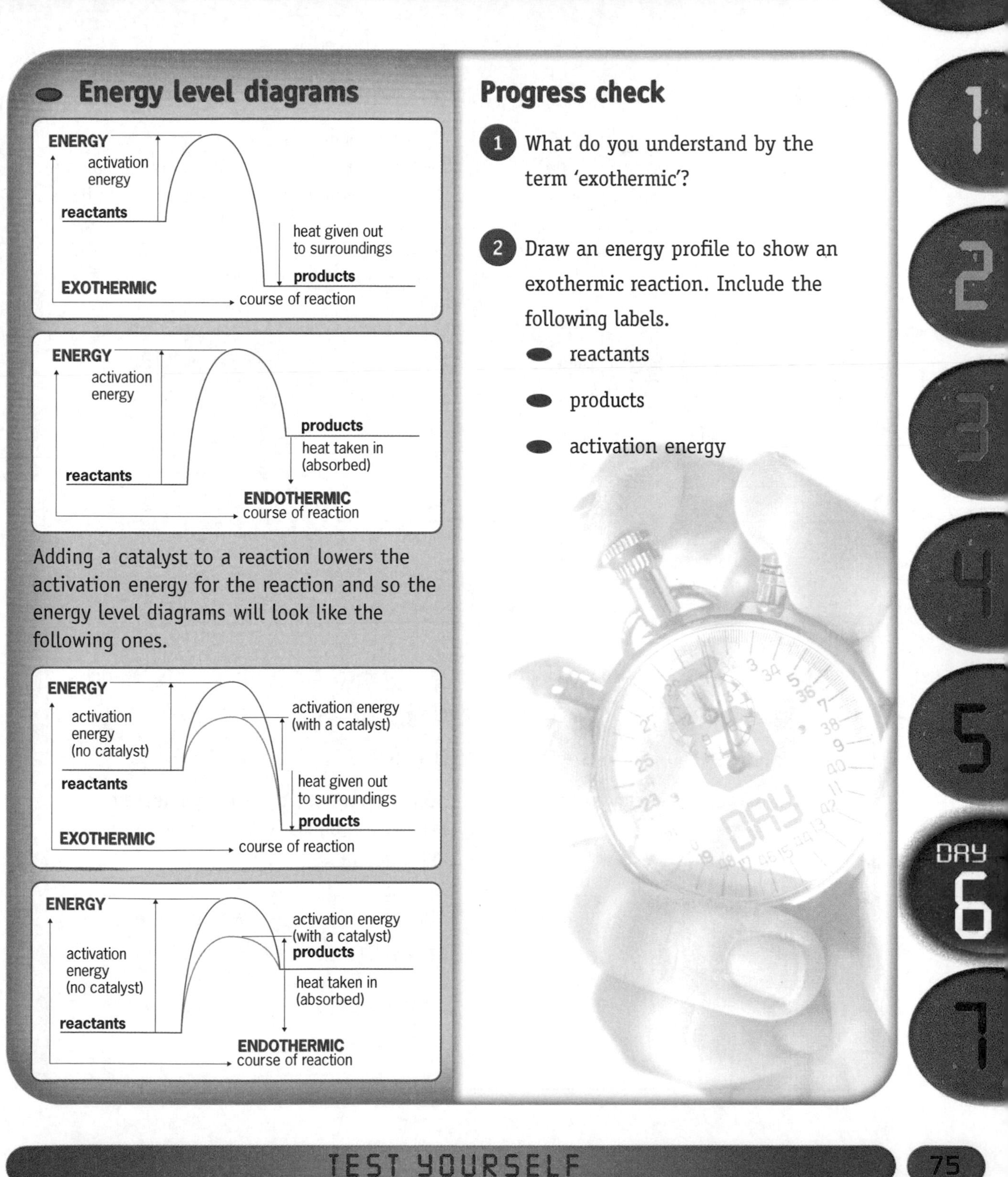

Adding a catalyst to a reaction lowers the activation energy for the reaction and so the energy level diagrams will look like the following ones.

Progress check

1. What do you understand by the term 'exothermic'?

2. Draw an energy profile to show an exothermic reaction. Include the following labels.
   - reactants
   - products
   - activation energy

CALCULATING ENERGY CHANGES

ΔH

The change in heat during a reaction is represented by the symbol ΔH. If the reaction is exothermic then ΔH has a negative value. If the reaction is endothermic then ΔH has a positive value.

Bond energy values

The bond energy is the amount of energy needed to break a particular bond. Breaking bonds is endothermic so bond energies have positive values. If a bond is made, this is an exothermic process so the bond energy is negative.

Bond	Bond energy ($kJ\,mol^{-1}$)	Bond	Bond energy ($kJ\,mol^{-1}$)
H–H	436	Br–Br	193
Cl–Cl	242	C–H	413
H–Cl	431	C–Cl	338
H–Br	366	C=O	743
O–H	463	O=O	496

Example 1

hydrogen + chlorine → hydrogen chloride

The symbol equation is: H_2 (g) + Cl_2 (g) → 2HCl (g)

but think of it as: H–H + Cl–Cl → H–Cl + H–Cl

Step 1: work out the energy needed to break the bonds in the reactant molecules.

Bonds broken	Energy (kJ mol–1)
H–H	436
Cl–Cl	242

Total energy in = 678 $kJ\,mol^{-1}$

Step 2: work out the energy released in making the bonds in the product molecules.

Bonds made	Energy ($kJ\,mol^{-1}$)
2 × H–Cl	2 × 431

Total energy out = 862 $kJ\,mol^{-1}$

You use bond energy values to work out how much energy is taken in to break the bonds in the reactant molecules and how much energy is given out in making the bonds in the product molecules. The difference in energy is known as ΔH (delta H).

Example 1 *continued*

Step 3: energy change (ΔH) = energy in – energy out

$\Delta H = 678 - 862 = -184\ \text{kJ mol}^{-1}$

The answer has a **negative value**. This means that the overall reaction is **exothermic**, i.e. more **energy is given out** in making new bonds than is used up in breaking old bonds.

Example 2

2H–Br → H–H + Br–Br

Bonds broken
Energy (kJ mol^{-1}):
2 × H–Br = 2 × 366
Total energy in = $732\ \text{kJ mol}^{-1}$

Bonds made
Energy (kJ mol^{-1}):
H–H = 436
Br–Br = 193

Total energy out = $629\ \text{kJ mol}^{-1}$

$\Delta H = 732 - 629 = +103\ \text{kJ mol}^{-1}$

The answer has a **positive value.** This means that the overall reaction is **endothermic**, i.e. less **energy** is given out in making new bonds than **is used up** in breaking old bonds.

Progress check

1. Calculate ΔH for the following reaction. (Use the bond energies in the table.)

   CH_4 (H–C–H with H above and below C) + 2O=O → O=C=O + 2H–O–H

2. Is the reaction exothermic or endothermic?

3. How do you know this?

4. The reaction of carbon with steam produces carbon monoxide and hydrogen. This reaction is endothermic. Explain, in terms of bond-making and bond-breaking, why this reaction is endothermic.

1 2 3 4 5 DAY 6 7

THE WATER CYCLE AND HARD/SOFT WATER

15 MINS

The water cycle

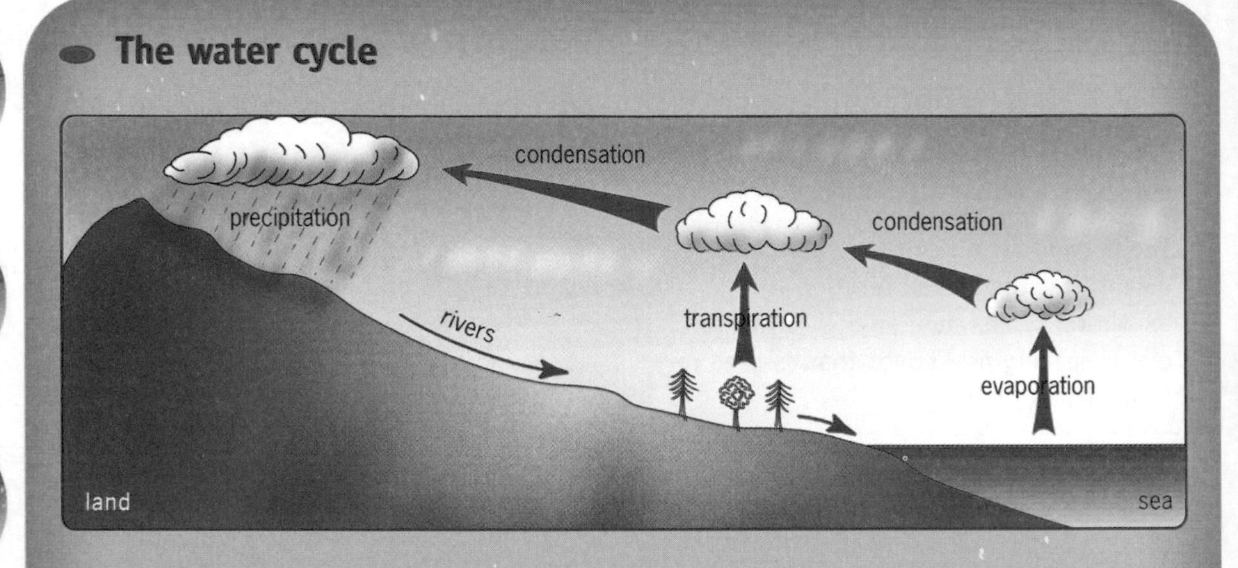

Hard water

When rain falls on certain rocks, e.g. limestone, it dissolves **calcium and magnesium compounds** present in the rock. These cause the water to be **hard**. Water without these ions dissolved in it is called **soft water**.

Hard water forms a **scum** when mixed with soap, **soft water** forms a smooth **lather**.

The soap test

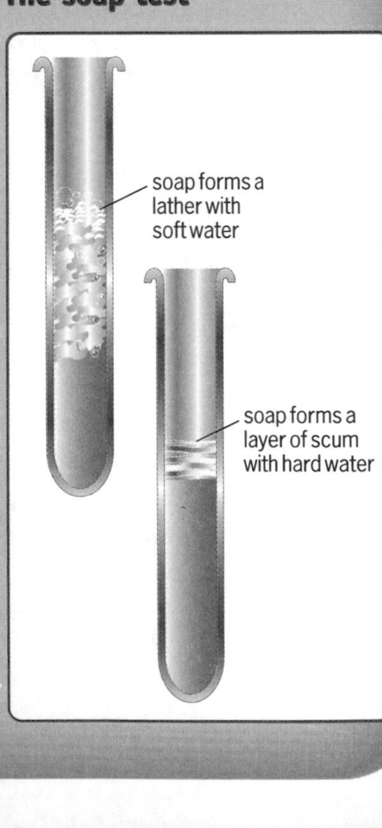

Permanent and temporary hardness in water

Hard water contains some, or all, of the following compounds:

- calcium sulphate
- calcium hydrogencarbonate
- magnesium sulphate
- magnesium hydrogencarbonate

An understanding of the water cycle helps you to track what happens to water in the environment. Water that passes through certain rocks dissolves certain ions and it becomes hard water.

15 MINS

Water that loses its hardness upon boiling is called **temporary hard water**. Temporary hardness is caused by the presence of calcium hydrogencarbonate. Hardness caused by other compounds is called **permanent hardness.**

The heat causes the calcium hydrogencarbonate to break down.

$Ca(HCO_3)_2 \rightarrow CaCO_3 + H_2O + CO_2$

The $CaCO_3$ forms a hard scale, often seen on kettle filaments and inside water pipes.

Advantages and disadvantages of hard water

- Dissolved calcium compounds are good for bones and teeth.
- The coating insides pipes can cause blockages.

Progress check

1. Explain how water in the ocean can end up falling as rain many miles away.
2. How can you tell the difference between hard and soft water?
3. What is temporary hard water?
4. Which substance causes temporary hardness in water?
5. Give one advantage and one disadvantage of hard water.

SOLUBILITY CURVES

Plotting a solubility curve

The table shows the maximum mass of copper(II) sulphate that can dissolve in 100 cm³ of water at different temperatures. When no more solute can dissolve in a solvent the solution is said to be **saturated**.

Temperature (°C)	10	20	30	40	50	60	70
Solubility (g/100 cm³ water)	27.5	32.0	37.8	44.6	53.2	61.8	72.8

This graph shows this information.

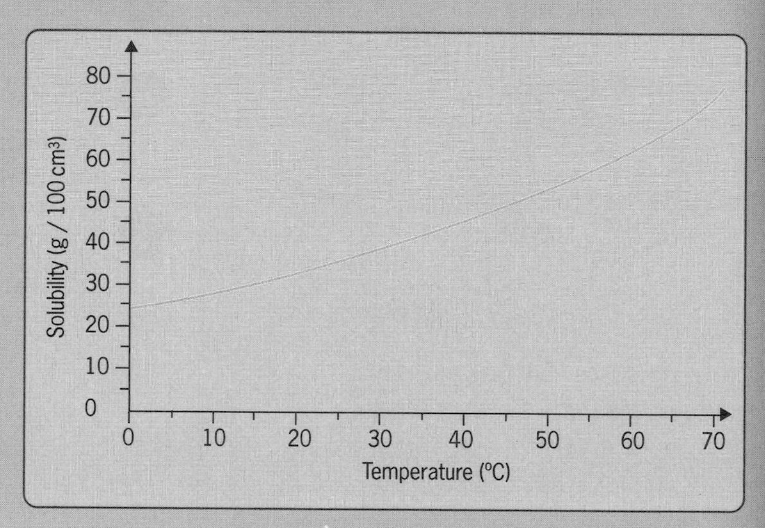

When a hot saturated solution cools, some of the solute will separate from the solution. This is called **crystallisation**.

You can use the solubility curve to estimate solubilities at different temperatures. For example, the solubility of copper(II) sulphate at 35°C can be estimated from the graph to be approximately 42 g/100 cm³.

The graph on the next page shows the solubility curves for sodium chloride and potassium nitrate.

The solubility of most solutes increases as the temperature increases. A plot of the maximum amount of solute that can be dissolved in a certain amount of solvent is called a solubility curve.

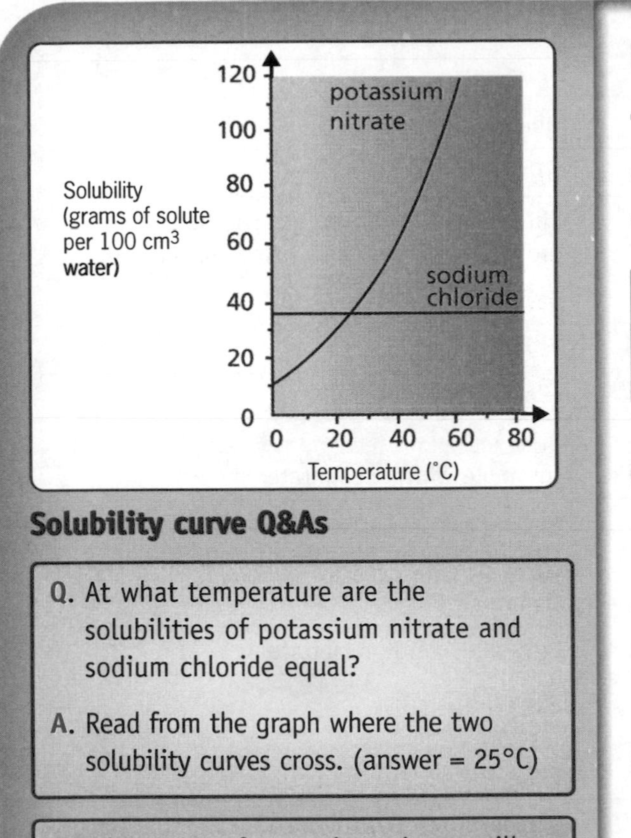

Solubility curve Q&As

Q. At what temperature are the solubilities of potassium nitrate and sodium chloride equal?

A. Read from the graph where the two solubility curves cross. (answer = 25°C)

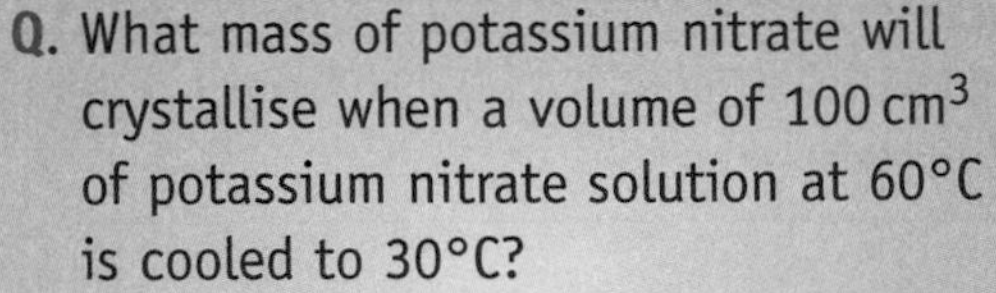

Q. What mass of potassium nitrate will crystallise when a volume of 100 cm^3 of potassium nitrate solution at 60°C is cooled to 30°C?

A. Work out the difference between the solubility at 60°C and 30°C. (answer = 80 g)

Q. How much more potassium nitrate can be dissolved at 45°C than at 35°C?

A. Work out the difference between the solubility at 45°C and at 35°C. (answer = 25 g)

Progress check

The table below shows the maximum mass of sodium nitrate that will dissolve in 100 cm^3 of water at different temperatures.

Temperature(°C)	0	10	20	30	40	50	60	70
Solubility (g/100 cm^3)	14	17	21	24	29	34	40	47

1. Plot a graph to show the solubility curve for sodium nitrate.

Use the graph to answer the following questions.

2. What is the solubility of sodium nitrate at 42°C?
3. Estimate the solubility of sodium nitrate at 80°C.
4. At what temperature is the solubility of sodium nitrate 45 g in 100 cm^3 of water?
5. What mass of sodium nitrate will crystallise if a saturated solution is cooled from 60°C to 20°C?

IDENTIFYING CATIONS

Tests for cations

Cation	Test	Observations
H^+	Use an indicator, e.g. UI	UI will turn red
	Any typical reaction of an acid, e.g. by reacting with a metal	With a metal hydrogen gas will be produced (which burns with a squeaky pop)
Li^+	Flame test	Dark red
Na^+	Flame test	Yellow
K^+	Flame test	Lilac
Ca^{2+}	Flame test Sodium hydroxide solution	Brick red Insoluble white precipitate
Ba^{2+}	Flame test	Apple green
Al^{3+}	Sodium hydroxide solution	White precipitate which dissolves in excess sodium hydroxide
Mg^{2+}	Sodium hydroxide solution	Insoluble white precipitate
Cu^{2+}	Flame test Sodium hydroxide solution	Green/blue Blue precipitate
Fe^{2+}	Sodium hydroxide solution	Green precipitate
Fe^{3+}	Sodium hydroxide solution	Orange/brown precipitate
NH_4^+	Warm with sodium hydroxide solution	Ammonia gas produced (pungent smell, turns UI paper blue)

Ionic equations

The reactions of some of the above cations with sodium hydroxide formed a precipitate. A **precipitation reaction** is one in which a solid forms when two solutions are mixed together.

An ionic equation can be written for the reaction of the cation with the hydroxide anion.

Example

$$Mg^{2+}(aq) + 2OH^-(aq) \rightarrow Mg(OH)_2(s)$$

$$Al^{3+}(aq) + 3OH^-(aq) \rightarrow Al(OH)_3(s)$$

The number of hydroxides added is the same as the number of positive charges the cation has.

Cations are atoms or groups of atoms that have lost at least one electron. There are a number of different chemical tests that can be carried out in order to identify which cation is present in a compound.

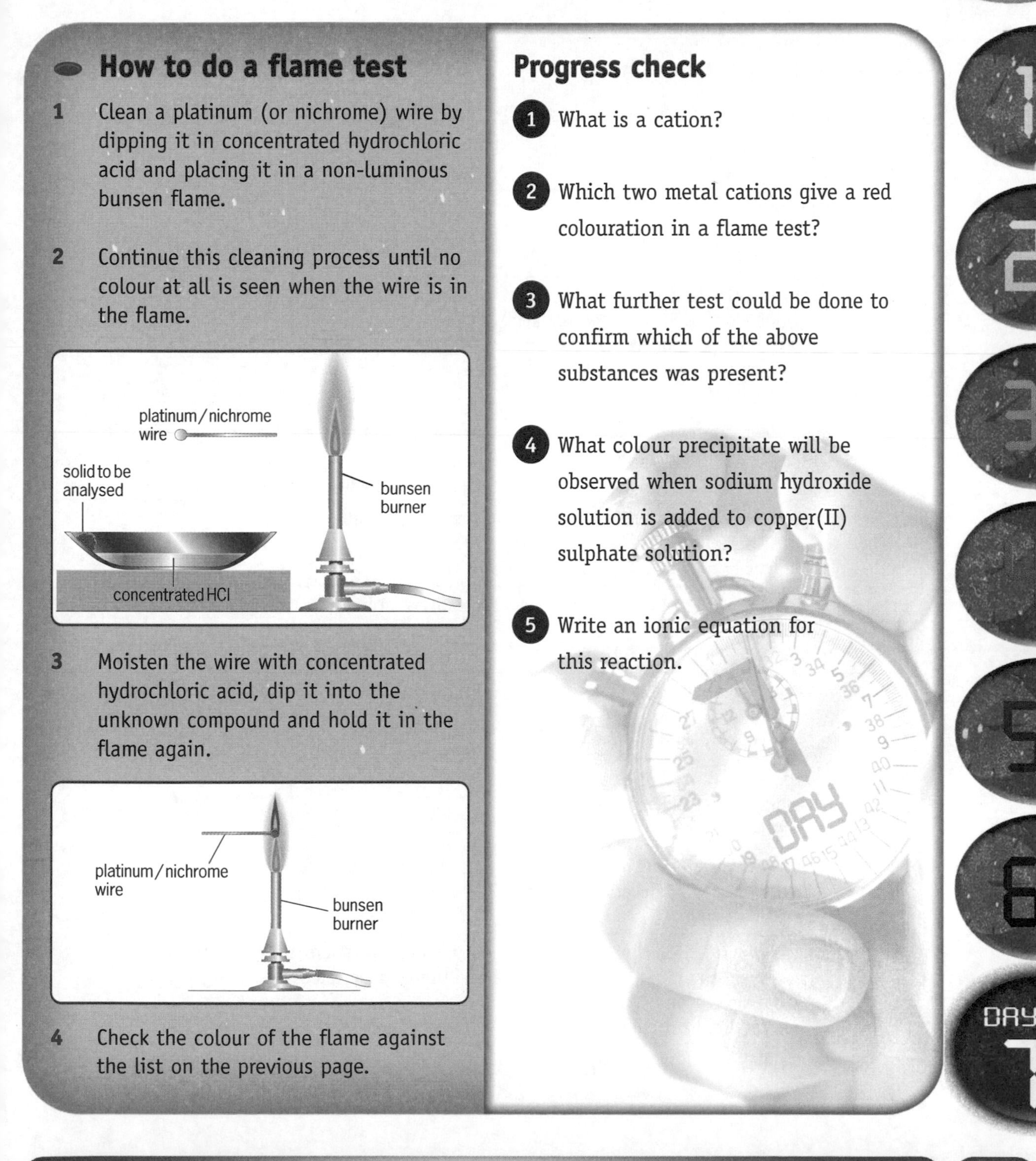

How to do a flame test

1. Clean a platinum (or nichrome) wire by dipping it in concentrated hydrochloric acid and placing it in a non-luminous bunsen flame.

2. Continue this cleaning process until no colour at all is seen when the wire is in the flame.

3. Moisten the wire with concentrated hydrochloric acid, dip it into the unknown compound and hold it in the flame again.

4. Check the colour of the flame against the list on the previous page.

Progress check

1. What is a cation?

2. Which two metal cations give a red colouration in a flame test?

3. What further test could be done to confirm which of the above substances was present?

4. What colour precipitate will be observed when sodium hydroxide solution is added to copper(II) sulphate solution?

5. Write an ionic equation for this reaction.

IDENTIFYING ANIONS AND INSTRUMENTAL ANALYSIS

Tests for anions

When testing for anions it may be necessary to dissolve the unknown substance in water first.

Anion	Test	Observations
OH^-	Add UI Heat with an ammonium salt	Turns blue A pungent gas is produced that turns red litmus paper blue
CO_3^{2-}	1. Add a small amount of dilute hydrochloric acid (or any other acid) 2. Pass the gas produced through limewater Copper carbonate and zinc carbonate can be identified by a colour change when heated	The limewater turns milky Copper(II) carbonate turns from green to black Zinc carbonate turns from white to yellow (the yellow turns to white upon cooling)
Cl^-	1. Add a small amount of dilute nitric acid 2. Add silver nitrate solution	White precipitate
Br^-	1. Add a small amount of dilute nitric acid 2. Add silver nitrate solution	Cream precipitate
I^-	1. Add a small amount of dilute nitric acid 2. Add silver nitrate solution	Yellow precipitate
SO_4^{2-}	1. Add a small amount of dilute hydrochloric acid 2. Add barium chloride solution	White precipitate
SO_3^{2-}	Heat with dilute hydrochloric acid and test the gas produced with moist blue litmus paper	Litmus paper turns red
NO_3^-	1. Add a small amount of sodium hydroxide solution 1. Add a small amount of aluminium powder and heat gently	A pungent gas is produced that turns red litmus paper blue

Anions are negatively-charged atoms (or groups of atoms). There are a number of different chemical tests that can be used to identify the presence of anions.

Instrumental analysis

As well as the above chemical tests (generally referred to as wet tests) elements and compounds can also be detected and identified by means of a variety of instrumental methods. Some instrumental methods, e.g. **flame photometry**, are suited to identifying elements while other instrumental methods, e.g. **mass spectrometry**, are more suited to identifying compounds.

Instrumental methods are accurate, sensitive (i.e. can work with minute quantities) and rapid. The development of modern instrumental methods has been aided by rapid progress in technologies such as computing and electronics.

However, machines can be expensive and complex, and need highly-skilled people to operate them.

Progress check

1. Describe the chemical test you could do to identify the anion present in a solution of sodium iodide.
2. How could you chemically distinguish between copper(II) carbonate and zinc carbonate?
3. What would you observe when barium chloride solution is added to a solution of magnesium sulphate?
4. Name one instrumental method of identifying the elements present in a substance.
5. Give one advantage and one disadvantage of this method compared to a chemical method of detection.

COLLECTION AND IDENTIFICATION OF GASES

Methods of collecting gases

Method 1 *Upward delivery*

This method is used to collect gases that are **less dense than air**, e.g. ammonia (NH_3).

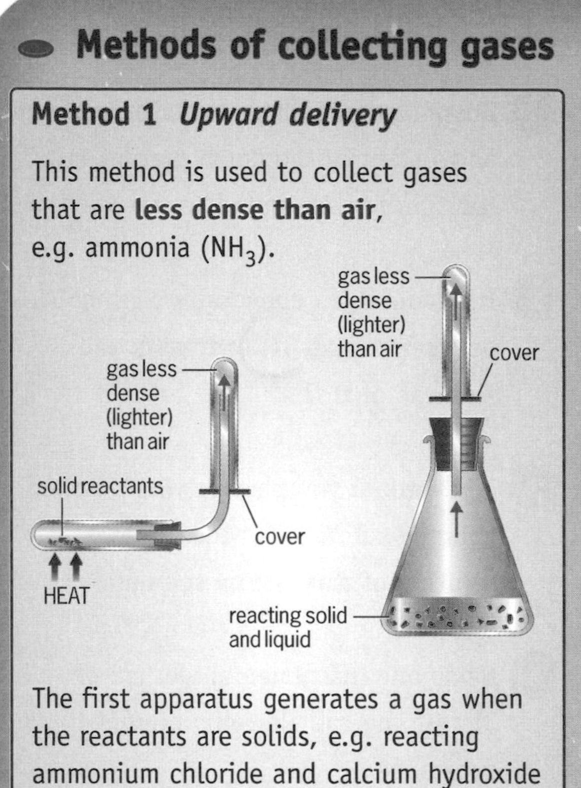

The first apparatus generates a gas when the reactants are solids, e.g. reacting ammonium chloride and calcium hydroxide to produce ammonia gas.

The second apparatus is more suited when one of the reactants is a liquid, e.g. reacting an acid with a metal to produce hydrogen gas.

Method 2 *Downward delivery*

This method is used to collect gases that are **more dense than air**, e.g. carbon dioxide (CO_2).

reactants
HEAT
cover
gas more dense (heavier) than air
reacting solid and liquid
cover
gas more dense (heavier) than air

Method 3 *Collection over water*

This method is used to collect any gas that is **not** (significantly) **soluble in water**, e.g. hydrogen (H_2).

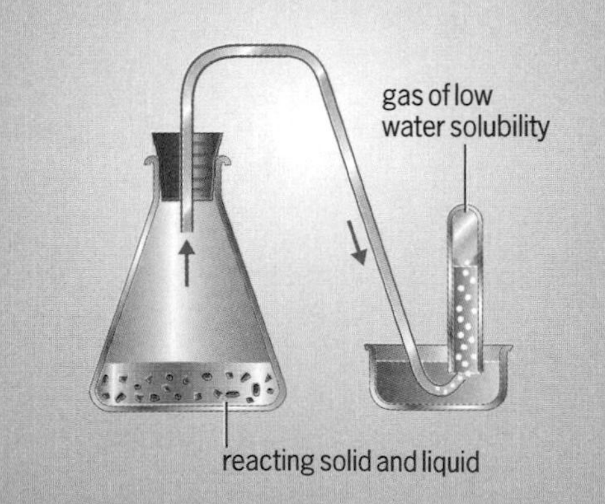

Gases can be collected in different ways and different chemical tests can determine their identity.

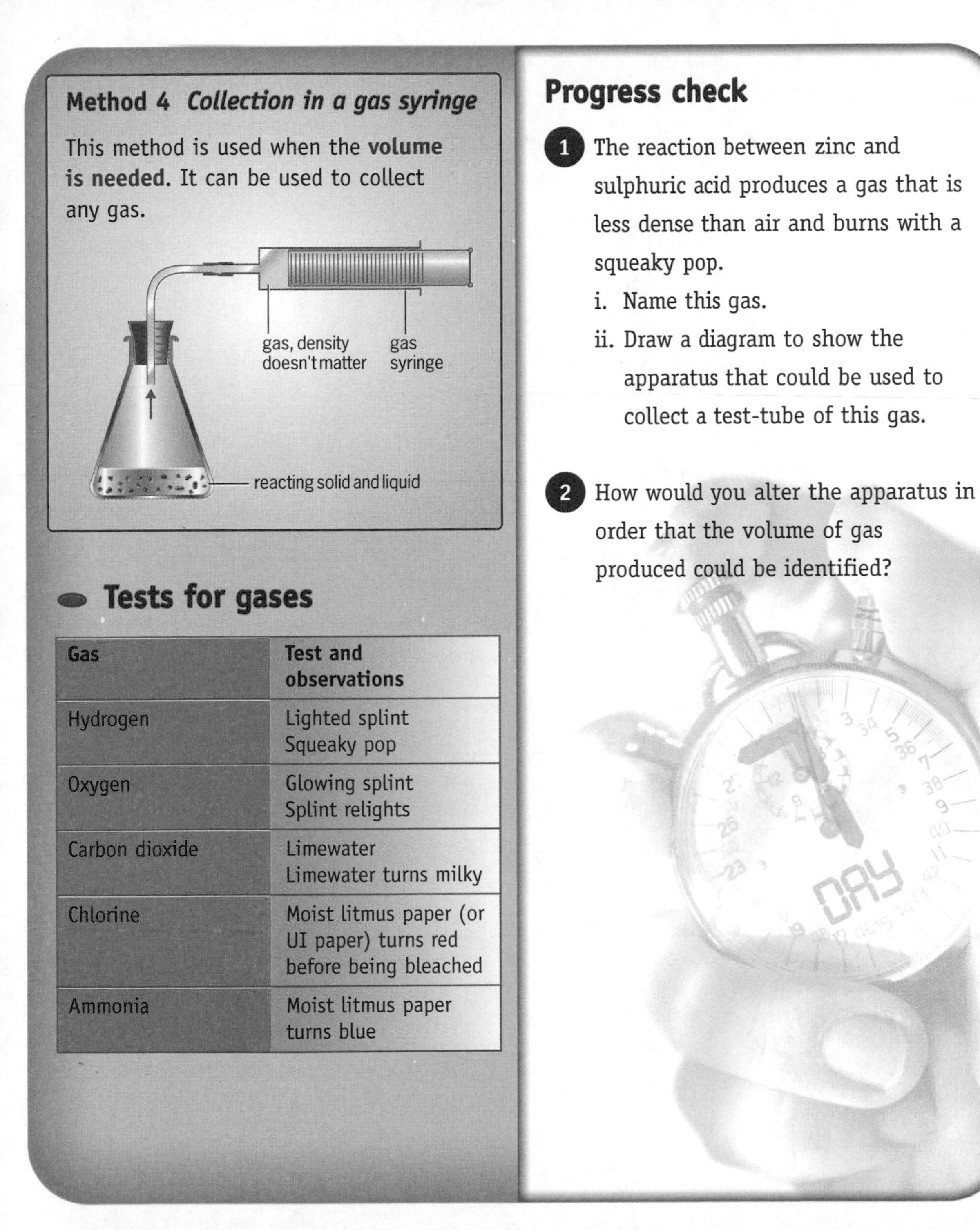

Method 4 *Collection in a gas syringe*

This method is used when the **volume is needed**. It can be used to collect any gas.

Tests for gases

Gas	Test and observations
Hydrogen	Lighted splint Squeaky pop
Oxygen	Glowing splint Splint relights
Carbon dioxide	Limewater Limewater turns milky
Chlorine	Moist litmus paper (or UI paper) turns red before being bleached
Ammonia	Moist litmus paper turns blue

Progress check

1. The reaction between zinc and sulphuric acid produces a gas that is less dense than air and burns with a squeaky pop.
   i. Name this gas.
   ii. Draw a diagram to show the apparatus that could be used to collect a test-tube of this gas.

2. How would you alter the apparatus in order that the volume of gas produced could be identified?

SULPHURIC ACID PRODUCTION

Uses of sulphuric acid

The pie chart represents the major uses of sulphuric acid.

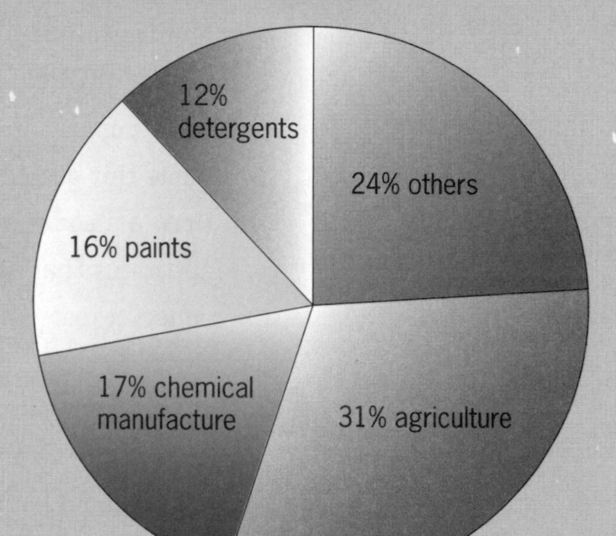

The contact process

The production of sulphuric acid occurs in three different stages.

The combustion of sulphur to form sulphur dioxide

$S + O_2 \rightarrow SO_2$

The oxidation of sulphur dioxide to form sulphur trioxide

$2SO_2 + O_2 \rightleftharpoons 2SO_3$

The sulphur dioxide formed in the first stage is mixed with more air and passed over a catalyst (vanadium pentoxide – V_2O_5). A temperature of approximately 450°C and a pressure of 1–2 atmospheres are used.

Sulphuric acid (H_2SO_4) is an industrially important chemical. It is used for a variety of different chemical processes. Sulphuric acid is manufactured by the contact process.

The conversion of sulphur trioxide into sulphuric acid

The sulphur trioxide is dissolved in concentrated sulphuric acid to form 'fuming sulphuric acid' or **oleum**.

$$SO_3 + H_2SO_4 \rightarrow H_2S_2O_7$$

The oleum is carefully diluted with water to make sulphuric acid again.

$$H_2S_2O_7 + H_2O \rightarrow 2H_2SO_4$$

Once there was a chemist .....

...A chemist he is no more ....

RIP
Prof.
Calvin
A. Mity

....For what he thought was H_2O .....

...Was H_2SO_4!!

ANSWERS

Atomic structure 1

1 b
2 negligible
3 positive (+1)
4 mass number – 63
atomic number – 29
5 $^{65}_{29}Cu$

Atomic structure 2

1 2
2 8
3 2,8,6
4

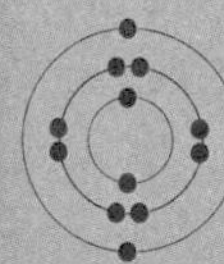

5 0
6 Atoms that have the same number of protons but different numbers of neutrons.
7 $\frac{(10 \times 20) + (11 \times 80)}{100}$
$= 10.8$

Covalent bonding and structures

1 b
2 simple covalent
3

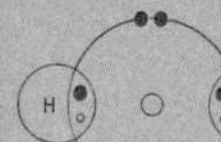

4 Hydrogen chloride is a simple covalent substance and it has a low melting point because there are weak forces between the molecules.
5 Hydrogen chloride does not conduct electricity because there are no free ions or electrons.
6 Diamond and graphite (giant covalent compounds) have high boiling points because there are lots of strong covalent bonds that require a lot of energy to break them.

Metallic bonding

1 Metal atoms lose their outer shell electrons. The metallic bond is the attraction between the resultant cations and the 'lost' electrons.
2 The 'sea of electrons' refers to the free moving electrons that surround the metal cations.
3 Conduct heat, conduct electricity, sonorous, malleable, ductile.
4 Metals can conduct electricity because the electrons are free moving.
5

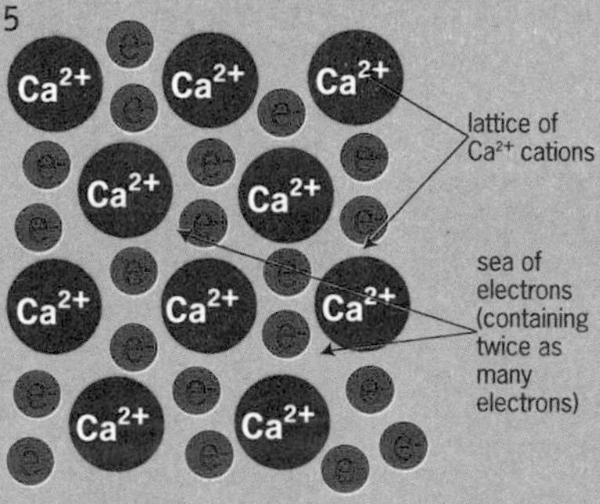

Crude oil: formation and separation

1 Crude oil was formed by the long-term effects of heat and pressure on decaying organic material in the absence of air.
2 A hydrocarbon is a compound containing carbon and hydrogen only.
3 Crude oil is separated by fractional distillation.
4

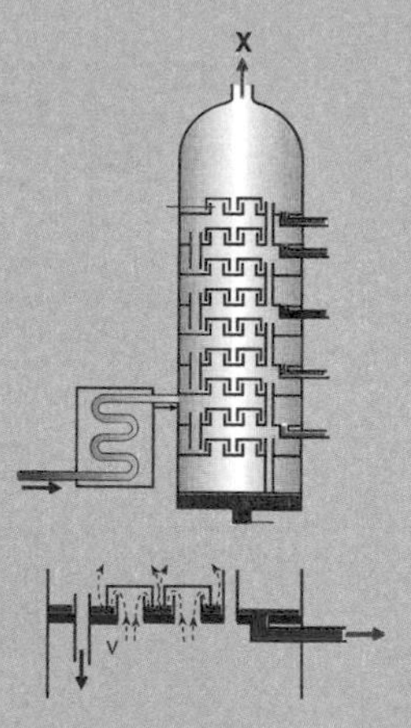

5 Hydrocarbons are separated according to their boiling point (or size of their molecule).

Alkanes and cracking

1 A saturated hydrocarbon is a compound of carbon and hydrogen that does not have any double bonds.
2 C_3H_8
3 ethane
4 heat and a catalyst
5 C_4H_{10}

Alkenes and polymerisation

1 Unsaturated means that there is a double bond between two carbon atoms in a compound.
2

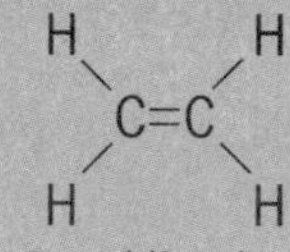

3 By adding bromine water. The octene (alkene) would decolourise it. The octane (alkane) would not.
4 Polymerisation is the process of joining many small molecules each containing a double bond (monomers) to form a long chain.
5

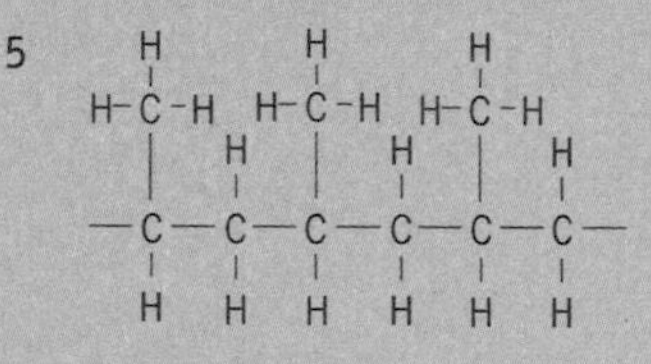

Isomerism

1 Isomers are compounds that have the same chemical formula but different chemical structures.

2

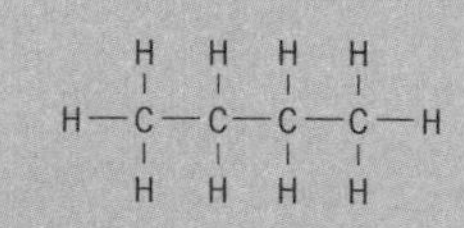

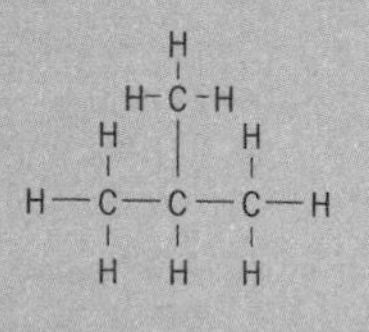

3

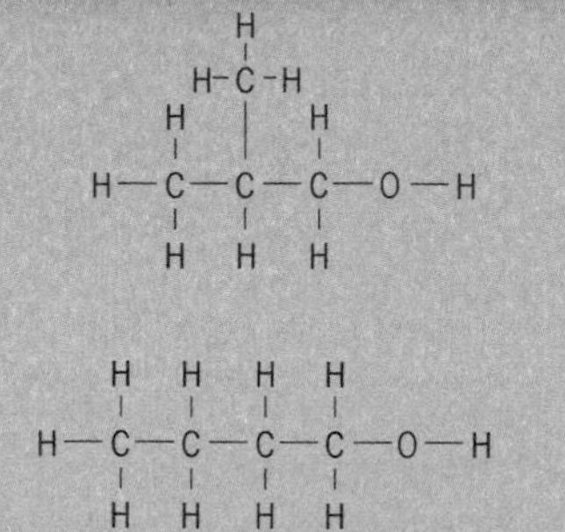

4 The first molecule (butane) will have the higher boiling point because there is less branching in the molecule, hence the molecules can get closer together, hence there will be stronger forces between the molecules.

5

```
         H
         |
       H-C-H
    H    |    H    H
    |    |    |    |
H — C — C — C — C — O — H
    |    |    |    |
    H    H    H    H
```

Alcohols and carboxylic acids

1 an OH group, R-OH

2

```
  H H
  | |
H-C-C-O-H
  | |
  H H
```

3 ethanoic acid

4 ethyl ethanoate (an ester) and water

5 The sulphuric acid acts as a catalyst.

6 e.g. ethanol as a solvent

7 e.g. ethanoic acid in making esters

The reactivity series and displacement reactions

1 any metal above Ca in the reactivity series, e.g. sodium

2 any metal below Pb in the reactivity series, e.g. copper

3 Cu is below H in the reactivity series.

4 It is very low in the reactivity series and therefore does not readily react with oxygen/water in the air.

5 A displacement reaction occurs when a more reactive metal takes the place of a less reactive metal in a compound.

6 magnesium + zinc sulphate → magnesium sulphate + zinc

7 Zinc is less reactive than magnesium and therefore it is unable to displace magnesium from a compound.

Iron and steel

1 iron ore (haematite), coke and limestone

2 The purpose of the limestone is to remove any acidic impurities in the iron ore (mainly sand – silicon dioxide).

3

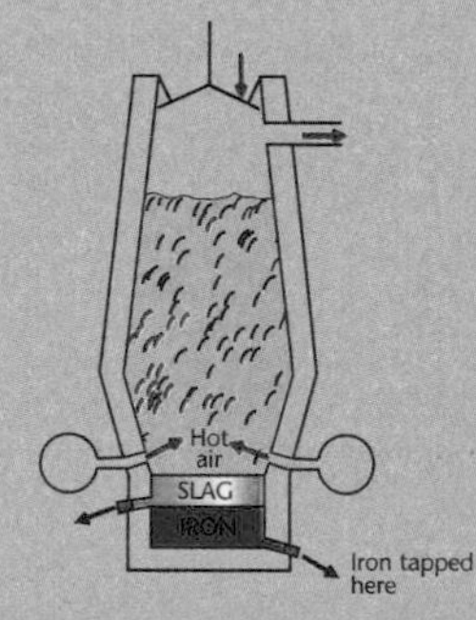

4 carbon monoxide

5 By removing the carbon in the steel by forcing oxygen through the molten iron. Different steels are then made by adding a metal such as chromium or nickel to it.

6 c

Extraction of aluminium and anodising

1 Aluminium is a very reactive metal and needs a powerful method of reduction to extract it from its oxide. Electrolysis is a more powerful method of reduction than the blast furnace.

2 bauxite.

3 $Al^{3+} + 3e^- \rightarrow Al$

4 reduction

5 Electrons are being gained (by the Al^{3+} ion).

6 The anode is made from graphite and this reacts with the oxygen gas that is formed at the anode producing carbon dioxide, hence the solid anode is slowly turning into a gas, so the anode needs to be regularly replaced.

7 Aluminium is a reactive metal and as soon as it is exposed to the air it reacts with oxygen to form a protective layer of aluminium oxide. This protects the surface of the aluminium and makes it appear less reactive than it actually is.

8 Anodising artificially thickens the oxide coating on aluminium further protecting it from reaction with oxygen and water in the atmosphere.

9 aluminium + oxygen → aluminium oxide

Copper purification and titanium production

1 Copper needs to be purified to remove any impurities as these reduce its electrical conductivity (i.e. increase its resistance).

2

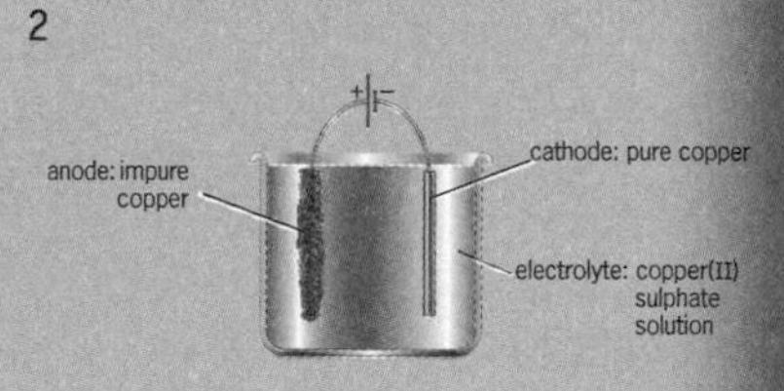

3 $Cu^{2+} + 2e^- \rightarrow Cu$

4 titanium(IV) chloride + magnesium → magnesium chloride + titanium

5 The oxygen in the air would react with the magnesium metal.

Pollution and useful products from limestone

1 CO_2; H_2O

2 water, sand and crushed rock

3 sand

4 See table on page 32.

Reversible reactions and equilibrium

1. A reversible reaction is one that can go forwards and backwards.
2. $\rightleftharpoons$
3. Equilibrium exists when the rate of the forward reaction is the same as the rate of the reverse reaction.
4. If the temperature is increased the yield of an exothermic reaction at equilibrium decreases.

Manufacture of ammonia

1. In the manufacture of ammonia a temperature of 450°C and a pressure of 200 atmospheres are used.
2. The temperature is a compromise temperature. A higher temperature would result in a lower yield of ammonia (because a high temperature favours the reverse reaction in an exothermic reaction). A lower temperature would mean that the rate of reaction is too slow to make the process economically viable. A higher pressure than 200 atmospheres would produce more ammonia but it would not be economically viable. It would also significantly increase the risk of an explosion.
3. Ammonia is used to make nitric acid.
4. The yield decreases.
5. The yield increases.
6. A lower temperature would mean the rate of production is uneconomical.

Working out formulae

1. Na_2O
2. $MgCl_2$
3. CS_2 (The formula C_2S_4 can be cancelled down to CS_2.)
4. Na_2CO_3
5. $LiNO_3$
6. $Al(OH)_3$

Balancing equations

1. $2Ca + O_2 \rightarrow 2CaO$
2. $2Cr + 6HCl \rightarrow 2CrCl_3 + 3H_2$
3. $CH_4 + 4Cl_2 \rightarrow CCl_4 + 4HCl$
4. $4Na + O_2 \rightarrow 2Na_2O$
5. $2P + 3Br_2 \rightarrow 2PBr_3$
6. $4Al + 3O_2 \rightarrow 2Al_2O_3$

Mole calculations 1

1. i. 40
   ii. 98
   iii. 158
2. i. 40%
   ii. 33% (nearest 1%)
   iii. 29% (nearest 1%)
3. i. Li_2O
   ii. $SiCl_4$

Mole calculations 2

1. 13.875 g
2. 7.75 g
3. 3.18 g
4. 2400 cm^3

Titrations 1

1. pipette, conical flask, indicator, burette, colour change
2. i. 0.025
   ii. 0.8 mol dm^{-3}

Titrations 2

1. $\frac{22.4 \times 0.1}{1000} = 0.00224$
2. 1:1
3. 0.002 24
4. 0.0896 mol dm^{-3}

Atmospheric chemistry

1. c
2. e.g. methane
3. The evolution of green plants allowed oxygen to be present in the atmosphere (oxygen is a by-product of photosynthesis).
4. By respiration and combustion of fossil fuels.
5. By photosynthesis and dissolving in the oceans.

The rock cycle

1. igneous, sedimentary and metamorphic
2. Intrusive igneous rocks have larger crystals than extrusive igneous rocks.
3. The rate at which the rock has cooled and solidified. Intrusive igneous rocks cool more slowly than extrusive igneous rocks.
4. Biological – plant root growth
   Chemical – acid rain
   Physical – freeze thaw
5. heat and pressure

The periodic table and the transition elements

1. by atomic number
2. periods
3. Elements are arranged in groups because they have similar chemical and physical properties. This is because they have the same numbers of electrons in their outer shells.
4. Transition metals are less reactive, denser, harder, have higher melting points than group 1 elements, coloured compounds.
5. e.g. iron as a catalyst (in the Haber process)

Non-metals: groups 7 and 0

1. e.g. non-conductors of electricity, low melting/boiling points, brittle/crumbly when solid
2. i. true
   ii. false
   iii. false
   iv. true
   v. false
3. Neon is an unreactive gas because neon atoms have a full outer shell of electrons. This means the atoms do not wish to lose/gain or share any electrons and hence they do not react with any other substance.
4. sodium iodide + bromine $\rightarrow$ sodium bromide + iodine
5. Bromine is more reactive than iodine because bromine atoms are smaller, hence the outer shell is closer to the nucleus hence it is easier for bromine atoms to gain an extra electron than iodine atoms.

The electrolysis of brine

1 hydrogen, chlorine and sodium hydroxide

2

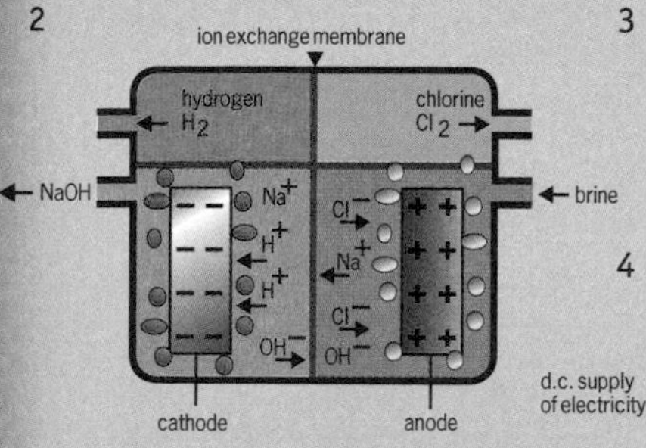

3 The chemical test for chlorine gas is to add moist blue litmus paper. The chlorine will turn the litmus paper red before bleaching it (turning it white).

4 Hydrogen – margarine manufacture
Chlorine – manufacture of bleach
Sodium hydroxide – manufacture of soap

5 Titanium electrodes are used because titanium is an unreactive metal and hence it will not react with any of the substances produced during this process and Ti conducts electricity.

Acids 1

1 green

2 7

3 i. hydrochloric acid + lithium hydroxide → lithium chloride + water
ii. sulphuric acid + calcium hydroxide → calcium sulphate + water
iii. nitric acid + sodium hydroxide → sodium nitrate + water

4 i. hydrochloric acid + potassium hydroxide
ii. sulphuric acid + lithium hydroxide
iii. nitric acid + calcium hydroxide

Acids 2

1 H^+

2 $H^+ + OH^- \rightarrow H_2O$

3 i. magnesium sulphate + hydrogen
ii. calcium oxide (or calcium hydroxide)
iii. sulphuric acid + copper(II) carbonate

4 Add sufficient zinc to the sulphuric acid until there is no more effervescence (fizzing and bubbling). Filter the mixture to remove any excess zinc. Warm the mixture to evaporate off some of the water. Leave the mixture in a warm place to allow the remaining water to evaporate/ the crystals to grow.

Rates of reaction 1

1 e.g. increasing the temperature, concentration, adding a catalyst, increasing the pressure (if the reactants are gases)

2 A catalyst lowers the activation energy of the reaction (by allowing the reaction to proceed by an alternative pathway).

3 As the reaction proceeds the concentration of the acid decreases.

Rates of reaction 2

1 a sufficient amount of energy called the activation energy.

2 Increasing the temperature
- means the particles will move more quickly, meaning more collisions
- means more particles will have the activation energy hence there will be more successful collisions.

3 If the concentration is doubled, then there will be twice as many reactant particles (in the same volume of liquid) hence there should be twice as many collisions hence the rate of reaction should theoretically double.

4 Smaller marble chips have a greater surface area than larger marble chips.

Reactions involving enzymes

1 sugar → ethanol + carbon dioxide

2 At temperatures above 45°C enzymes start to denature making them less effective.

3 See table on page 72.

4 See table on page 72.

Exothermic and endothermic reactions

1 Exothermic means 'heat is given out'.

2

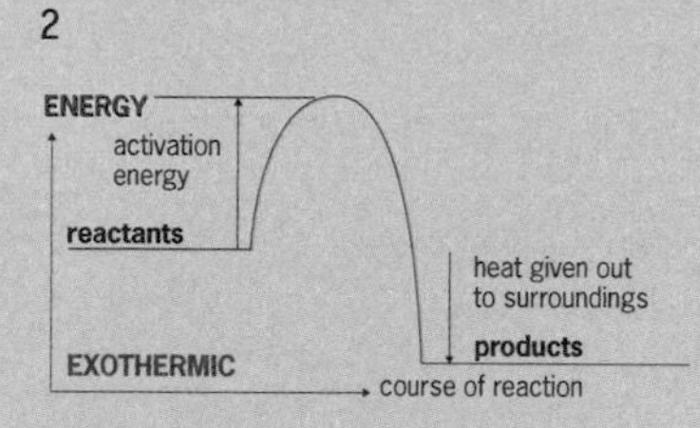

Calculating energy changes

1 Bonds broken
4 × C–H 4 × 413 = 1652
2 × O=O 2 × 496 = 992
1652 + 992 = 2644 kJ mol^{-1}
Bonds formed
2 × C=O 2 × 743 = 1486
4 × O–H 4 × 463 = 1852
1486 + 1852 = 3338 kJ mol^{-1}
ΔH = bonds broken – bonds formed =
2644 – 3338 = –694 kJ mol^{-1}

2 The reaction is exothermic.

3 Exothermic reactions have a negative ΔH value.

4 In an endothermic reaction more energy is used in breaking the bonds in the reactant molecules than is released in making the new bonds in the product molecules.

The water cycle and hard/soft water

1. Water evaporates from the oceans, condenses and forms as clouds and these move.
2. Hard water will form a scum with soap whereas soft water forms a lather.
3. Temporary hard water is water that loses its hardness upon boiling.
4. The hydrogencarbonate (HCO_3^-) ion causes temporary hardness.
5. See list on page 79.

Solubility curves

1

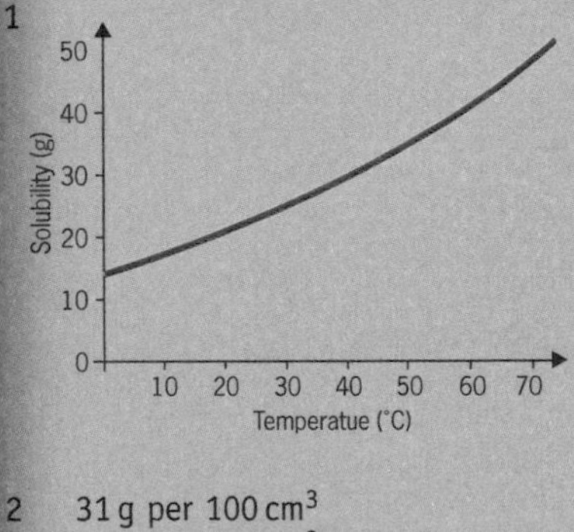

2. 31 g per 100 cm^3
3. 60 g per 100 cm^3
4. 66°C
5. 19 g

Identifying cations

1. A cation is a positively-charged ion (an atom that has lost electrons in order to gain a full outer shell).
2. calcium and lithium
3. Add sodium hydroxide solution: Ca^{2+} would form an insoluble white precipitate in the presence of hydroxide ions.
4. a blue precipitate
5. $Cu^{2+}(aq) + 2OH^-(aq) \rightarrow Cu(OH)_2(s)$

Identifying anions and instrumental analysis

1. Add a small amount of dilute nitric acid followed by silver nitrate solution. A yellow precipitate will be observed.
2. By heating both substances. Copper(II) carbonate will turn from green to black. Zinc carbonate turns from white to yellow.
3. an insoluble white precipitate
4. See page 85.
5. See page 85.

Collection and identification of gases

1. i. hydrogen
   ii.

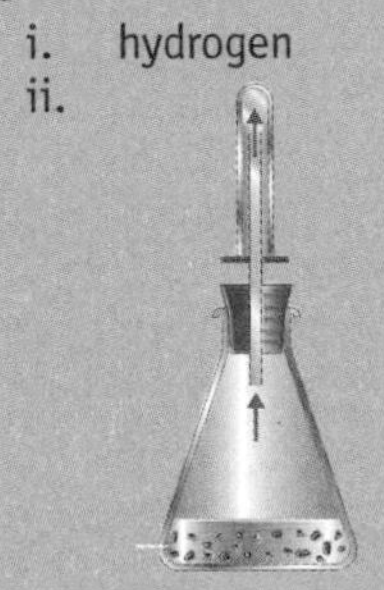

2. Collect the evolved gas in a gas syringe.